The Labyrinth of Desertification

Jaime Martínez-Valderrama · Jorge Olcina Cantos
Editors

The Labyrinth of Desertification

Sixteen Case Studies in Spain

GOBIERNO DE ESPAÑA
VICEPRESIDENCIA TERCERA DEL GOBIERNO
MINISTERIO PARA LA TRANSICIÓN ECOLÓGICA Y EL RETO DEMOGRÁFICO
Fundación Biodiversidad
Plan de Recuperación, Transformación y Resiliencia
Financiado por la Unión Europea
NextGenerationEU

Springer

Editors
Jaime Martínez-Valderrama
Estación Experimental de Zonas Áridas (EEZA)
CSIC
Almería, Spain

Jorge Olcina Cantos
Universidad de Alicante
Alicante, Spain

ISBN 978-3-032-09647-0 ISBN 978-3-032-09648-7 (eBook)
https://doi.org/10.1007/978-3-032-09648-7

This work was supported by the Biodiversity Foundation of the Ministry for the Ecological Transition and the Demographic Challenge (MITECO) of the Government of Spain, within the framework of the Recovery, Transformation and Resilience Plan (PRTR), funded by the European Union - NextGenerationEU.

This Springer imprint is published by the registered company Springer Nature Switzerland AG
The registered company address is: Gewerbestrasse 11, 6330 Cham, Switzerlande

If disposing of this product, please recycle the paper.

Preface

In 2027, it will be 100 years since the term desertification was first used. It was coined by the French forestry engineer Louis Lavauden (1927) in the context of colonial Africa to describe the transformation of productive Tunisian lands into deserts as a result of human activity in the tropical forest regions of Africa. This colonial perspective blamed native populations for deforesting the landscape and promoting the advance of the desert. However, the nomadic pastoralism they practiced was, in fact, the most effective strategy for surviving in these meager environments, where ochre was the natural color of the landscape.

Nonetheless, the tendency to "green" the land through irrigation or reforestation—under the belief that this halts desertification—persists to this day (Fundación Nueva Cultura del Agua 2020). Likewise, stereotypical images of sand dunes and cracked soils continue to be linked to desertification, reinforcing the idea that desert advance or droughts are the root cause of the problem.

Over the years, and despite ongoing research, interpretations of desertification have varied widely. In a recent essay, Reynolds (2021) ironically highlighted some of the most outlandish academic misuses of the term. For instance, one might hear statements such as "In Ukraine, soil degradation is 38% compaction, 4% salinization, and also 0.2% desertification"—a claim that overlooks the difficulty in measuring the problem and contradicts the established hierarchy, where desertification encompasses soil degradation, which in turn includes salinization. Or: "In northern China, desertification is the primary form of land degradation because it causes soil degradation." Such examples illustrate the deep conceptual confusion surrounding the term.

Global change has further distorted the concept. On one hand, excluding hyper-arid zones from consideration seems outdated, as modern technologies and infrastructure now enable their exploitation—growing crops in deserts with deep groundwater is no longer extraordinary. On the other hand, the delocalization of production and consumption displaces environmental degradation, often beyond arid zones, challenging the geographical and conceptual boundaries of desertification. A striking example is the deforestation of South American forests to produce soy for Mediterranean livestock feed (Martínez-Valderrama et al. 2021).

Although desertification has been clearly defined—land degradation in arid, semi-arid, and dry sub-humid areas resulting from various factors, including climatic variations and human activities (UNCCD 1994)—it remains widely misunderstood, both in public discourse and academic circles. It continues to be shaped by subjective interpretations and hampered by the lack of standardized methodologies for its measurement. Even the UNCCD itself has used misleading imagery to raise awareness about the issue (Prince and Podwojewski 2019), inadvertently reinforcing the widespread myth that desertification simply equates to the expansion of deserts (Martínez-Valderrama 2024). The latest *World Atlas of Desertification* (Cherlet et al. 2018) exemplifies this conceptual ambiguity: paradoxically, it does not include a single map of desertification.

The high degree of subjectivity involved in identifying when we are truly facing a process of desertification—and the fact that it does not correspond to a single, easily measurable phenomenon—have traditionally hindered efforts to monitor and assess it effectively. This book presents 16 case studies from Spain to explore the diverse nuances of landscapes and situations commonly associated with desertification.

As illustrated in Figure 1, desertification can be represented as a labyrinth, in which finding the exit—understood here as successfully identifying when a landscape is undergoing desertification, whether incipient, active, or in a stable and seemingly irreversible phase—is particularly challenging. Through the 16 selected case studies, our aim is to gradually explore a spectrum of situations: from those that clearly do not constitute desertification, to those that may or may not evolve into it, and finally, to those that unambiguously align with the official definition. However, this clarity—and the classification proposed—is not universally accepted. When we add this confusion to the conceptual ambiguities previously noted by Reynolds, the metaphor of the labyrinth becomes especially fitting. Thus, the primary goal of this book is to foster understanding as a necessary prerequisite for addressing desertification effectively. Escaping the labyrinth, therefore, means laying the groundwork for the development of coherent and effective solutions.

To introduce the discussion, Chap. 1, "*A Framework of Conceptual Principles for Addressing Desertification*" presents a brief essay that outlines a set of principles a case must meet to be considered an instance of desertification. The subsequent case studies serve either to challenge these criteria or to demonstrate why certain situations cannot be classified as desertification. The typology includes three categories: cases that do not constitute desertification, those that may or may not evolve into it, and those that clearly fall within its definition. However, such clarity is not always universally shared—hence the use of the maze metaphor. This metaphor aptly reflects the book's central objective: to promote a deeper understanding of desertification as a prerequisite for addressing the issue effectively. Escaping the maze, in this sense, involves establishing the conceptual and methodological foundations necessary to develop coherent and actionable solutions.

The first three cases examine situations that, while potentially related to desertification, do not in themselves represent desertification processes. Chapter 2, "*Droughts, Aridity and Desertification: The Climatic Factor*," explores the significance of arid conditions and droughts in creating a context conducive to desertification. As discussed in the definitional framework, aridity constitutes the climatic background in which

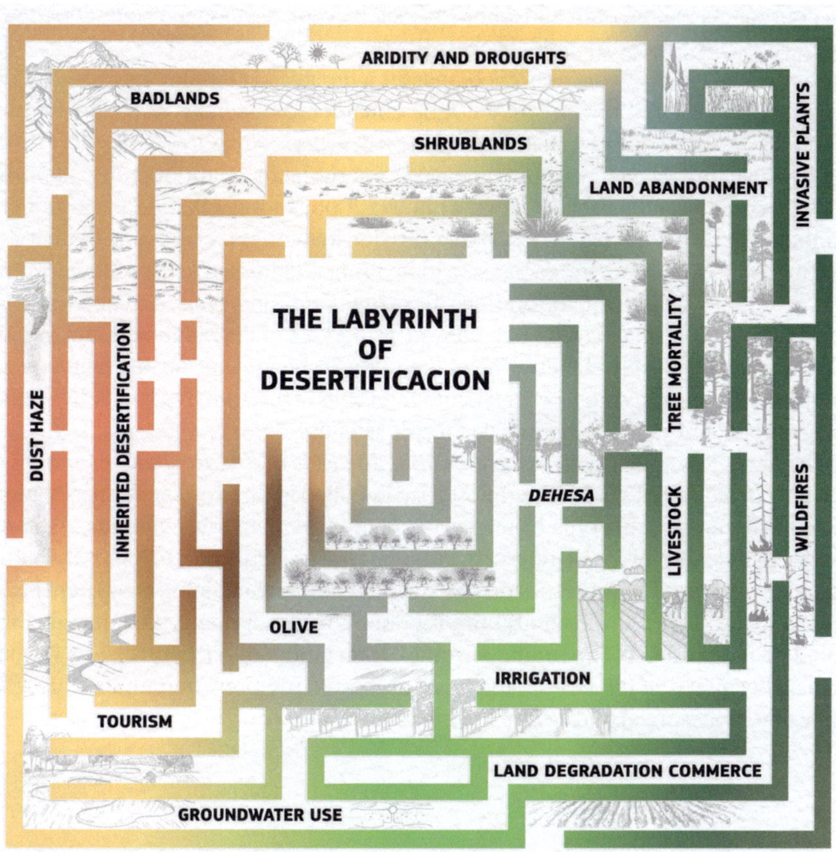

Fig. 1 Within the labyrinth of desertification, we find cases that society often perceives as desertified—such as badlands—yet technically are not, and others that give an impression of development and prosperity—such as irrigated agriculture—but may ultimately lead to irreversible land degradation

desertification can occur, while droughts are among the climatic variations capable of triggering it. However, neither factor alone results in land degradation.

Chapter 3, "*Badlands: What Desertification Is Not*," addresses one of the most persistent misconceptions surrounding desertification: its confusion with geomorphological landscapes, such as badlands, where human activity has played no role in their formation. Indeed, the very term "badlands" already conveys the difficulty of traversing or utilizing these areas, thereby excluding the anthropogenic component that is central to the definition of desertification.

Chapter 4, "*The Harms and Benefits of Dust*," further clarifies that Saharan dust and haze events affecting the Iberian Peninsula do not constitute desertification—despite their negative impacts on public health—nor does the Sahara "invade" southern Europe. However, when dust emissions originate from agricultural activities—specifically as a consequence of erosion and soil loss—such events can indeed be understood as manifestations of desertification processes.

The second typology of cases encompasses a broad and complex spectrum of gray areas—situations that may or may not be considered instances of desertification. Chapter 5, "*Shrub Encroachment in Drylands: Scientific Basis Regarding Its Positive and Detrimental Effects*," explores the role of shrublands in arid regions. Identified as a "desertification landscape" (Fig. 2) in the 2008 National Action Programme to Combat Desertification (MAGRAMA 2008) due to their associations with increased wildfire risk—by altering fire regimes (a topic further addressed in the following chapter)—as well as with the deterioration of grazing lands and even rural abandonment (see Chap. 12), shrublands are often viewed as markers of degradation. However, this chapter reveals that shrublands may, in many cases, be the only viable land cover in certain territories, playing a crucial ecological role in arid environments by sequestering carbon and functioning as true "fertile islands."

Chapter 6, "*Wildfires and Desertification: The Role of Fire Regime*," examines the changing dynamics of wildfire regimes. Also included in the aforementioned National Action Programme, this type of landscape comprises numerous variants—some of which can indeed evolve into desertification processes, while others represent natural ecological regeneration processes that play a vital role in maintaining ecosystem health. It is the disruption of these natural fire regimes—driven by poor land management, global change impacts (including shrub encroachment and the spread of invasive species, discussed in the next chapter), or excessive fire prevention measures—that can lead to widespread and severe land degradation.

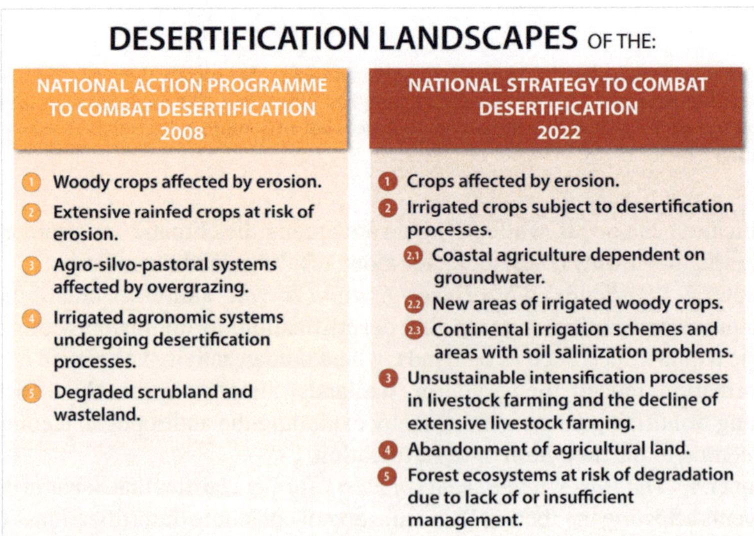

Fig. 2 Desertification landscapes in Spain

Chapters 7, "*Invasive Plants as a Symptom and Vehicle of Degradation*," and 8 "*Tree Mortality Processes Accelerate Desertification*," examine degradation dynamics not traditionally linked to desertification. Chapter 8 is explicitly addressed in the National Strategy to Combat Desertification—which replaced the 2008 Action Programme (MITERD 2022)—while Chap. 7, although less commonly associated with desertification, relates to both wildfire dynamics and biodiversity loss, in addition to causing economic losses within the agricultural sector.

The case study analysis continues with two chapters focused on livestock farming. Chapter 9, "*Moving Livestock to Reverse Desertification Landscapes*," emphasizes the importance of livestock mobility as a key management tool to mitigate the potential threats that pastoralism may pose in relation to desertification. On one hand, the prolonged presence of livestock in a fixed location can trigger processes such as overgrazing, erosion, and a qualitative decline in pastureland. On the other hand, undergrazing is associated with shrub encroachment (as discussed in Chap. 5) and an increased risk of wildfires (see Chap. 6). Herbivory plays a crucial role in reducing flammable biomass in forests, while also representing a potentially valuable economic activity in depopulated or abandoned rural areas—territories often linked to desertification, as explored in Chap. 11.

The *Dehesa*—the subject of Chap. 10, "*The Dehesa: A Cultural Landscape Facing Future Challenges*"—has historically been classified as a desertification-prone landscape due to overgrazing. However, with the phasing out of Common Agricultural Policy (CAP) incentives that once promoted livestock intensification, this agrosilvopastoral system has increasingly been recognized as one of the most sustainable landscape types. Its long-term viability depends, among other factors, on the livestock mobility discussed in the previous chapter.

Chapter 11, "*Desertification by Land Abandonment*," presents one of the most controversial cases in the volume. Rather than being driven by overexploitation of natural resources, this form of degradation results from underuse of the land. Rural depopulation and land abandonment give rise to various forms of environmental degradation, including shrub encroachment and the emergence of landscapes prone to fire—both phenomena discussed in earlier chapters. However, some perspectives argue that abandonment marks the beginning of landscape rewilding, a process fundamentally opposed to desertification.

Finally, the second typology of cases concludes with Chap. 12, which focuses on tourism (*Tourism and Water Stress: A Worrying Convergence in Time and Place*). Although the connection to desertification may seem indirect, a closer look at the specific timing and locations of tourist activity reveals a significant overlap with periods and areas of peak water demand. This convergence places extraordinary pressure on already limited water resources, potentially pushing certain ecosystems beyond their thresholds and contributing decisively to the depletion of water reserves—thereby accelerating processes of desertification.

The third group of case studies is the most directly related to desertification and includes landscapes identified in both the National Action Programme to Combat Desertification (PAND) and the more recent national strategy (see Fig. 2). The first of these, discussed in Chap. 13, "*Inherited Desertification: Mining, Deforestation and*

Abandonment in the Sierra de Gádor Mountains (Almería)," presents a situation that is difficult to classify as active desertification. These landscapes were functionally productive until the relatively recent past—just a couple of centuries ago, a mere instant on the geological timescale—but collapsed after a brief period of resource exploitation and have yet to recover. This case illustrates one of the most severe dimensions of the problem: its irreversibility.

Chapters 14–16 examine the most classic desertification landscapes. Chapter 14, "*Irrigation as a Source of Development and Degradation*," explores the consequences of water resource overexploitation. While irrigation has historically been a key land use in Spain's drylands, its uncontrolled expansion—driven largely by profitability—has placed intense pressure on both surface and groundwater resources. This has led to their quantitative and qualitative degradation, as well as the deterioration of wetlands associated with these water bodies.

Woody crops are closely linked to soil erosion and, more recently, to the degradation of water resources. Among them, olive grove—one of Spain's most emblematic crops—is the focus of Chap. 15, "*Desertification of the Olive Grove: A Complex Challenge for a Diverse Agricultural System*." Traditionally a rainfed system, olive groves were once comparable to forested landscapes in terms of their associated biodiversity. However, their intensification—again driven by economic profitability—has triggered severe erosion, particularly when olives are planted on steep slopes with no ground cover. The gradual shift toward irrigated systems has further integrated olive cultivation into other desertification-prone landscapes, traditionally occupied by herbaceous or horticultural crops.

Chapter 16, "*Hiding Desertification with Technology: Irrigation and Water Imbalance in Southern Spain*," delves into the deterioration of aquifers, which were once reliable water reserves and should have been key allies in the fight against climate change. However, their overexploitation has led to the emergence of striking green patches in some of Spain's most arid regions, creating a misleading image of prosperity and abundance. Around the use of these strategic water reserves, agro-industrial models have flourished—shielded by the rhetoric of technical efficiency—yet they ultimately undermine territorial sustainability and exacerbate social inequality.

Finally, Chap. 17, "*The Global Trade of Degradation Distorts the Scope of Desertification*," addresses a phenomenon characteristic of the current era of global change: the decoupling of production and consumption zones. This raises a critical dilemma—who should be held accountable for the environmental degradation caused by the production of raw materials and food in one region that are ultimately consumed in another? The chapter questions whether desertification is, in effect, being exported when local resources are spared at the expense of degrading distant ecosystems, or whether consumer countries are absolved of responsibility simply by purchasing those goods. What is clear is that the environmental accounting of this degradation must be properly assigned; otherwise, we risk creating blind spots that undermine efforts to neutralize desertification.

References

Cherlet M, Hutchinson C, Reynolds J et al (2018) World atlas of desertification. Publication Office of the European Union, Luxembourg

Fundación Nueva Cultura del Agua (2020) Falacia 4. El regadío es la solución para frenar la desertificación. In: Desmontando falacias sobre agua y cambio climático. https://fnca.eu/desmontandofalacias/falacia-4/. Accessed 20 Oct 2021

Lavauden L (1927) Les Forêts du Sahara. Rev des Eaux Forêts 65:265–277

MAGRAMA (2008) Programa de Acción Nacional contra la Desertificación. Madrid. Ministerio de Agricultura y Medio Ambiente, Madrid, Spain

Martínez-Valderrama J (2024) Desertificación, cuando el territorio hace aguas [Desertification: lands in troubled waters]. Tirant lo Blanch, Valencia, España

Martínez-Valderrama J, Sanjuán ME, Barrio G et al (2021) Mediterranean landscape re-greening at the expense of South American agricultural expansion. Land 10:204. https://doi.org/10.3390/land10020204

MITERD (2022) Estrategia Nacional de Lucha contra la Desertificación en España. Madrid, Spain

Prince SD, Podwojewski P (2019) Desertification: inappropriate images lead to inappropriate actions. L Degrad Dev 31:677–682. https://doi.org/10.1002/ldr.3436

Reynolds JF (2021) Desertification is a prisoner of history: an essay on why young scientists should care. Ecosistemas 30:2302. https://doi.org/10.7818/ECOS.2302

UNCCD (United Nations Convention to Combat Desertification) (1994) United Nations Convention to Combat Desertification in those countries experiencing serious drought and/or desertification, particularly in Africa. Paris, France

Almería, Spain — Jaime Martínez-Valderrama
Alicante, Spain — Jorge Olcina Cantos

The original version of the book has been revised. A correction to this book can be found at https://doi.org/10.1007/978-3-032-09648-7_19

Prologue

Desertification is one of the most misunderstood environmental challenges of our time. Often portrayed in oversimplified terms—frequently equated with nothing more than soil erosion or drought—it has long been addressed with a narrow set of tools and concepts. But desertification, particularly in Mediterranean drylands such as those of Spain, is a multifaceted and deeply contextual phenomenon. It is shaped by the interplay of biophysical processes, land-use decisions, economic pressures, and cultural legacies. To tackle it effectively, we must move beyond classical approaches and embrace a broader, more integrative perspective.

This book, *The Labyrinth of Desertification: Sixteen Case Studies in Spain*, is a response to that need. Through a series of carefully chosen case studies across diverse Spanish landscapes—from arid southeastern basins to forested mountain ranges and touristic coastal zones—it offers a grounded and nuanced examination of the many faces of land degradation and desertification in Spain. Its central aim is to challenge dominant narratives, question established diagnostic tools, and highlight both the complexity and the context-dependence of desertification processes.

A core message emerging from the chapters in this volume is the central role of agriculture in driving land degradation. Agricultural practices—especially those that have intensified over recent decades such as irrigated- and greenhouse-based agriculture—have had profound impacts on soil health, vegetation structure, and hydrological dynamics. Irrigation schemes have transformed natural river regimes and depleted and polluted aquifers across Spain; mechanization and deep ploughing have accelerated erosion and reduced soil organic matter; the expansion of monocultures has simplified ecosystems and reduced their resilience. At the same time, the economic pressures driving such intensification—market demands, subsidy structures, and land consolidation—must be understood as part of the problem.

Yet, paradoxically, the abandonment of traditional agricultural landscapes has also contributed to new forms of degradation. In many regions, the decline of extensive farming and pastoralism has led to soil erosion, the loss of open habitats, and increased vulnerability to wildfires. These seemingly opposing processes—overuse and abandonment—can coexist even within the same territory, creating a dynamic and often contradictory mosaic of land-use change.

Livestock management occupies a particularly complex position within this narrative. For decades, grazing was often regarded as a degradation agent—blamed for overgrazing, compaction, and vegetation loss. While overstocking can indeed be damaging, a growing body of evidence now points to the potential of well-managed pastoralism to enhance biodiversity, control the amount of fuel, and maintain landscape heterogeneity. A chapter in this volume shed light on how pastoralism, when adapted to local conditions and knowledge systems, can act as a powerful tool for sustaining ecosystem services and preventing land degradation. This calls for a re-evaluation of livestock not as a problem to be removed, but as a management ally to be better understood.

The book also brings into focus several key drivers of desertification that have not traditionally received the attention they deserve. One is the role of tourism, particularly in coastal and insular environments. The chapter on tourism reveals how massive urban development—driven by seasonal demand and speculation—has led to irreversible land cover changes, water overextraction, and habitat fragmentation, among other problems. In these cases, the degradation is not always visible as soil loss but is manifested through the disruption of hydrological cycles, vegetation transformation, and the collapse of ecological buffers.

Wildfires are another central concern, especially as climate change increases their frequency and intensity. The loss of traditional fire management practices, coupled with the accumulation of biomass in abandoned rural areas, has created conditions ripe for large, high-severity fires. Fire-induced desertification is not just about the immediate loss of vegetation; it includes long-term changes in soil structure, water repellency, and post-fire erosion risks. Importantly, the social dimension of wildfire risk—linked to depopulation, land-use change, and forest policy—must also be accounted for when addressing this key environmental issue.

Land abandonment, meanwhile, is examined not simply as a retreat from the land but as a profound transformation of socio-ecological systems. While abandonment can lead to spontaneous rewilding or carbon sequestration in some contexts, it can also result in a loss of cultural landscapes, decline of agrobiodiversity, and increased homogenization of vegetation. The implications for desertification are complex and context-specific, requiring an analytical lens that can account for both ecological processes and socioeconomic trajectories.

What makes this book particularly valuable is its commitment to grounding theory in real-world cases. The chapters of the book are rooted in specific Spanish landscapes and are co-authored by scholars with deep knowledge of local ecosystems and their socioeconomical and ecological dynamics. The result is a rich tapestry of examples that defy simplistic classifications. The examples included in this book are a great reminder that indicators, models, and maps—however useful—can never fully capture the lived reality of landscapes undergoing change driven by human activities.

Another strength of this volume is its insistence on critical reflection. The authors do not simply present data; they interrogate the frameworks and assumptions behind desertification assessments. They challenge the idea that all degradation is visible, measurable, or even negative in the same way across contexts. They invite us to consider issues of scale—how processes that appear neutral or beneficial at one level may be harmful at another. And they call for new governance approaches that are participatory, adaptive, and place-based.

Taken together, the case studies in *The Labyrinth of Desertification* offer not just a diagnosis but also a way forward. They point to the importance of integrated land management, combining scientific insight with local knowledge. They advocate for policies that recognize complexity rather than simplify it, and that support multi-functional landscapes rather than chase single-sector efficiency. And above all, they emphasize the need to reconnect with the land—not only physically, but intellectually and culturally.

As we confront the global challenge of land degradation, Spain's drylands offer lessons of both caution and hope. This book captures those lessons with honesty, rigor, and clarity. It is my hope that it will serve as a valuable resource for researchers, land managers, students, and policymakers—not only in Spain but in drylands around the world. May it help us to better understand the maze of desertification, and to find paths toward more resilient and sustainable futures.

Thuwal, Saudi Arabia
June 2025

Fernando T. Maestre

Contents

Editor's Biography

Jaime Martínez Valderrama holds a Ph.D. in agricultural engineering from the Polytechnic University of Madrid. He specializes in desertification and global change, with a primary research focus on understanding the relationship between land use and its environmental impacts, particularly in arid regions. His work involves the use of integrated simulation models that link socioeconomic variables with their biophysical impact. He has published numerous articles in high-impact journals, and his research has been utilized by various public and private institutions. He is also dedicated to science communication, regularly writing popular science articles, contributing to various science outreach blogs, and serving as editor of *Arida Cutis*. Additionally, he is the author of *Deserts and Desertification and Desertification, Land in Troubled Waters*. Currently, he leads the ATLAS project, which aims to develop a desertification map for Spain and to compile all the information related to this problem.

Jorge Olcina Cantos (Alicante, 1966) obtained his degree in geography and history and a master's degree in land use planning from the University of Alicante. He is a Full Professor on regional geographic analysis at the University of Alicante, where he teaches subjects on land use planning, climatology, and natural hazards. He has focused his research on various geographical topics (climatology, natural hazards, land use planning, and theory of geography). He is the author or co-author of more than 200 publications (scientific papers, book chapters, and monographs). He has participated in several research projects on geographical and historical topics developed at the University of Alicante. He was an official speaker at the International Year of Planet Earth (2008), declared by UNESCO. He is a visiting professor at universities in Spain, Europe, and Latin America, and member of the Editorial Board of various scientific journals on geographical and environmental topics. He was an evaluator for the Fifth United Nations IPCC Report on Climate Change; president of the Spanish Geography Association (AGE) (2017–2021); and director of the Chair of Climate Change at the University of Alicante.

Chapter 1
A Framework of Conceptual Principles for Addressing Desertification

Gabriel del Barrio, María E. Sanjuán and Jaime Martínez-Valderrama

Abstract The term "desertification" is burdened by its colonial origin and later development based more on ad hoc effects descriptions than on proper understanding of the socio-ecological processes involved. It is ambiguous and cases are difficult to detect for two reasons. First, desertification is often tackled from a symptomatic, rather than systemic, viewpoint. Second, desertification is characterized using an array of attributes, a variable subset of which must be present, but none of them being decisive. In this essay, we propose a set of principles a case must fulfil to qualify as desertification. These are: (1) Desertification causes land degradation; (2) Prioritizing economic efficiency exclusively results in negative environmental and social externalities; (3) Desertification takes place in coupled but spatially separated land subsystems; (4) Desertification affects large geographical areas; (5) Resources origins and sinks in a desertification setup may be spatially telecoupled; (6) Early detecting of desertification requires to combine "slow" and "fast" variables of socio-ecosystems; (7) Abrupt land use changes can foster land degradation; (8) The perception of some "fast" socioeconomic variables can lead to a misperception of desertification; (9) Aquifers and their associated basins make coupled exploited subsystems of desertification; (10) Desertification risk is different from desertification condition; and (11) A main problem facing desertification monitoring and remediation is the

G. del Barrio (✉) · M. E. Sanjuán · J. Martínez-Valderrama
Estación Experimental de Zonas Áridas (EEZA), CSIC, Almería, Spain
e-mail: gabriel@eeza.csic.es

M. E. Sanjuán
e-mail: marieta@eeza.csic.es

J. Martínez-Valderrama
e-mail: jaimonides@eeza.csic.es

J. Martínez-Valderrama and J. Olcina Cantos (eds.), *The Labyrinth of Desertification*,
https://doi.org/10.1007/978-3-032-09648-7_1

desertification baseline. This proposal is timely in view of some recent frameworks that demand a precise and consistent characterization of desertification: Sustainable Development Goals in general, and particularly Land Degradation Neutrality, and Convergence of Evidence.

Keywords Land degradation · Land degradation neutrality · Land use change · Telecoupling · Systemic approach · Drylands

The Winding Path of Desertification

The term "desertification" was coined almost a century ago (Lavauden 1927). Since then, progress in addressing this issue has been limited (Grainger 2015; Safriel 2017; Martínez-Valderrama et al. 2020a). Despite the existence of a global convention dedicated to combating desertification, the ambiguity of the term has led to overlapping and sometimes conflicting actions and policies. Among the numerous definitions of desertification, the official one from the United Nations Convention to Combat Desertification (UNCCD) states it as "land degradation in arid, semi-arid, and dry sub-humid areas resulting from various factors, including climatic variations and human activities." In this context, degradation refers to "the reduction or loss of biological or economic productivity and land complexity" (UNCCD 1994).

Despite containing operative elements that define its scope, causes, and what is understood by degradation, the concept of desertification remains shrouded in confusion. Burdened by its colonial origins, desertification has become "a prisoner of history," as Reynolds (2021) describes in a recent essay. Consequently, there is no consensus on its extent or severity, which hampers mapping desertification. The latest World Atlas of Desertification (WAD; Cherlet et al. 2018) corroborates this stating that "Although 'desertification' remains in the title, this atlas represents a significant departure from the first two editions of the World Atlas of Desertification because deterministic maps on global land degradation are not presented."

Lack of conceptual consistency often results in proposed solutions that are nonsensical and counterproductive. A current example is the push for reforestation in drylands, which in many cases exacerbates the problem by straining the already scarce water resources of these areas (Mátyás and Sun 2014; Turner et al. 2023) and increasing plant density in naturally open ecosystems such as savannas (Bond 2019).

Much of this confusion stems from the term "land". Although the UNCCD specifies that it refers to "a terrestrial bioproductive system, which includes soil, water, vegetation, other biomass, as well as ecological and hydrological processes occurring within the system," in many situations, land is equated solely with soil, neglecting important processes and uses. As Meyfroidt et al. (2022) pose it, "land is the source and focus of multiple meanings and values" that are dynamic over time and influence claims about its expected use and benefits (Li 2014). Land is also culturally and symbolically laden, further complicating its interpretation and management.

This contributes to the ambiguity of the term "land degradation." Whilst the UNCCD definition clearly emphasizes its anthropogenic component, the inclusion of environmental and economic aspects introduces subjectivity, ultimately leading to value judgements. On one hand, there is robust biophysical and ecological knowledge that allows for scientific measurement of indicators of change in ecosystem functions. On the other hand, the interpretation of these physical measures in terms of their impact on land benefits is ultimately based on people's views and definitions, which can be broadly shared but also conflicting (Meyfroidt et al. 2022). Specific changes in land systems, such as soil erosion, groundwater degradation, and loss of organic matter, are typically included in definitions of land degradation and generally have overwhelmingly negative impacts on human societies. Further, definitions of land degradation often extend beyond these specific aspects. If addressing these issues translates into the cessation of economic activity, it can also be considered land degradation. The UNCCD itself reflects this as a duality in its approach to desertification: developing countries primarily view it as a developmental issue, while developed countries consider it an environmental one (Ortiz and Tang 2005).

Inconsistent definitions, differing criteria among member countries, and the multifaceted nature of desertification have diminished the influence and relevance of the UNCCD, especially when compared to its sister initiatives Convention on Biological Diversity (CBD) and the United Nations Framework Convention on Climate Change (UNFCC). This is evident in the significantly lower number of academic articles on desertification and land degradation compared to the other two conventions, as well as the general lack of interest and awareness among the public (Reynolds 2021). The near invisibility of events like the United Nations Decade on Deserts and the Fight Against Desertification (2010–2020) and the International Year of Deserts and Desertification in 2006 further corroborates this point (Martínez-Valderrama 2024).

However, the distortions around the concept of desertification are even more worrying, as they affect the actions that can be taken to address the problem (Grainger 2015; Cherlet et al. 2018; Reynolds 2021; Sanjuán et al. 2022). We believe that there are two main reasons behind such distortions. First, desertification is often tackled from a symptomatic, rather than systemic, viewpoint. Degraded landscapes are described in physiognomic terms addressing vegetation cover or erosion landforms (Kosmas et al. 1999; Imeson 2011; Kuderna and Zsolnay 2014), rather than using ecological variables such as net primary productivity or turnover rate (Wessels et al. 2007; Puigdefabregas et al. 2009). As a result, desertification is treated as a landscape type rather than as an ecological state (del Barrio et al. 2021). This hinders linking environmental features of desertification to the socio-economic systems that caused them in the first place. The epitome of such a mix between causes and consequences is the indistinct use of "desertification" and "land degradation". The second reason is that desertification is commonly characterized using a fuzzy array of attributes, a subset of which must be present for a case to qualify as desertification, but none of them being decisive to this purpose. Technically, this form of classification is named polythetic (Needham 1975) as opposed to a monothetic classification, where each case must satisfy all of a fixed set of conditions to belong to a class.

Polythetic identification of desertification cases creates own problems. While the essence of desertification was already formulated and modelled some decades ago (Puigdefabregas 1995; Ibáñez et al. 2008b), the fact that it emerges in drylands worldwide under particular socio-economic setups has fostered the consideration of different ecological trajectories, which have been recounted as syndromes (Hill et al. 2008) or desertification landscapes (Martínez-Valderrama et al. 2022). This has the advantage of enabling solutions based on the local human population, but obscures theoretical advance with sometimes unnecessary details.

In this chapter, we propose a set of principles that must be fulfilled in order to qualify a case as desertification. The current state of the art of the topic does not lend well to a strict monothetic system without raising the abstraction to unpractical levels. However, each principle of the proposed set should be such that it can be used to tell desertification from other environmental pressures occurring in drylands.

This proposal is timely given the co-occurrence of novel frameworks that demand a precise and consistent characterization of desertification, which is currently lacking. First, we briefly outline these recent paradigms to frame the need of that characterization. The rest of the chapter sets out the desertification principles.

New Initiatives as Opportunities to Refine the Concept of Desertification

In recent years, some developments have emerged that can redirect the UNCCD's approach and establish guidelines for addressing the issue with greater success. The first is the Land Degradation Neutrality (LDN) initiative, which became formalized in 2015 as part of the Sustainable Development Goals (SDGs), specifically goal 15.3: *By 2030, combat desertification, restore degraded land and soil, including land affected by desertification, drought and floods, and strive to achieve a land degradation-neutral world* (UNCCD 2016). LDN recognizes that land degradation is inherent in human activity and proposes to compensate for it by following certain premises (Orr et al. 2017; Cowie et al. 2018). While the UNCCD's scope is limited to drylands, the LDN conceptual framework can be applied to all land types, uses, and ecosystem services. Prevention measures are prioritized over restoration actions in LDN guidelines, reflecting the irreversible nature of the problem.

The second innovative development revolves around the concept of Convergence of Evidence (CE). This approach aligns with the primary emphasis of LDN, which is prevention. It offers a diagnostic methodology aimed at effectively addressing this core objective. Though not entirely new (e.g., in Spain, the "desertification landscapes" were identified pioneering the underlying idea in the CE (Martínez-Valderrama et al. 2022), the CE is an alternative to desertification maps proposed by the European Commission in the last World Atlas of Desertification (WAD, Cherlet et al. 2018). Desertification is polythetic, as explained above, and multiscale geographical decouplings between its causes and effects makes mapping impossible. Moreover, assessing the state of ecosystems is often subject to biases and prejudices, which undermine the possibility of successful mapping (Gibbs and Salmon 2015; Prince 2016). Consequently, the best products available so far are thematic or monothetic maps, which either capture a particular aspect such as land degradation (del Barrio et al. 2016), or identify land syndromes responding homogeneously to corresponding arrays of desertification causes (Hill et al. 2008).

The CE, on the other hand, proposes to overlay layers of biophysical and socio-economic data to detect sites where economic development may trigger degradation processes that compromise sustainable development. Rather than being replicated in each pixel of the territory, the analysis should reveal hotspots prone to desertification and be *interpreted* in their socio-political context, giving local and regional conditions an important role in desertification processes.

As both LDN and CE are relatively recent and may represent a renaissance in the tackling of desertification, it is essential that the concepts are clear so that the proposals are effective. In the light of works that are appearing under the umbrella of these two paradigms, this does not seem to be the case. For example, in addressing LDN, land is still, in some cases, equated with soil (Albaladejo et al. 2021), leaving aside some of the most important issues related to desertification, such as the overexploitation of water resources.

The rationale of the CE, which considers local perceptions of desertification, implies the need to have a clear framework of what desertification is, so as not to include as such cases of degradation that do not fit a set of criteria. Reynolds (2021) exposes many of these inconsistencies with a note of humor. So, for example, we can find studies related to desertification that state that "land degradation is 38% compaction and 4% salinization, and it is also 0.2% desertification", or that, in northern China, desertification is the main form of land degradation, as it causes soil degradation. There is even work on land degradation and desertification in Iceland (Aradóttir et al. 2013), or that in tundra areas, thawing permafrost can lead to desertification (Wang et al. 2004).

Finally, desertification is intricately related to several Sustainable Development Goals (SDGs) (Martínez-Valderrama et al. 2023a). Although the term explicitly appears only in SDG 15 and target 15.3 ("By 2030, combat desertification, restore degraded land and soil, including land affected by desertification, drought, and floods, and strive to achieve a land degradation-neutral world"), there are numerous opportunities to address desertification through other goals as well. For instance, the socioeconomic dimension of desertification is closely tied to SDGs 1 (No Poverty) and 2 (Zero Hunger), which necessitate economic development actions rather than purely conservation efforts. The progressive decoupling between the causes of degradation and its manifestations is linked to the increasing urbanization of populations, which shifts their environmental footprint thousands of kilometers away. In this context, addressing desertification and land degradation is closely related to SDG 12 (Responsible Consumption and Production). These examples underscore the necessity of having a clear and unified understanding of desertification to streamline efforts effectively.

A Set of Principles for Detecting Desertification

In the following, we present a set of principles that distinguish desertification from other environmental problems that may also cause ecological degradation in drylands, for example, environmental pollution.

A conventional form to illustrate these principles would be to resort to examples from different countries or regions worldwide. However, while this might favor archetypical adherence to the principle under consideration, it would certainly add to the conceptual heterogeneity underlying desertification, which is the opposite of what is sought here.

Alternatively, all the examples have been drawn from a single country, seeking a territorial coherence that might have not been possible should them had been arbitrarily extracted from all over the world. With this choice the syndrome paradigm may evolve from its current state, according to which desertification occurs in syndromes associated to a country or region, to a more refined state where a territory prone to desertification is likely to produce a variety of concurrent syndromes. This will help creating a robust base for the Convergence of Evidence approach.

The cases selected for this purpose are based in Spain, a country where footprints of global change events associated with desertification are varied both in time and in space. Their relationship with the desertification principles is shown in Table 1.1, and details of it can be found in the comprehensive descriptions in later chapters of this book.

1. *Desertification causes land degradation.* It is paramount to clarify the link between desertification and land degradation. The former is a root cause of the latter. Desertification comprehend a process driven by socio-economic factors that encourage the unsustainable use of slow-renewing resources in vulnerable environments with poor water balance (MEA 2005), i.e., the system reaches its resilience threshold (Puigdefabregas 1995). These drivers can be technological, political or economic, and they often act in synergy. The Dust Bowl, for example, was the result of the interaction of multiple factors (Lockeretz 1978; Schubert et al. 2004), including the invention of the deep plough, the development of new cold-resistant wheat varieties, settler expansion policies toward the west of the continent, and a market niche that emerged after the collapse of Russian wheat exports due to the Bolshevik revolution. Land degradation, on the other hand, refers to strictly environmental variables, and is the ecological process whereby ecosystems lose both biomass and net primary productivity as a consequence of simplification linked to overexploitation (del Barrio et al. 2016). Whilst their causal relationship is always maintained, loss of management options associated with land degradation activates positive feedback –of explosive behavior– in advanced stages of desertification: the more degraded a land becomes, the more the local population is forced to insist on the same land use system that is generating desertification.
2. *Aiming solely at economic efficiency results in negative environmental and social externalities.* The shift from isolated, developing societies with self-sufficient

Table 1.1 Desertification case studies described as respective chapters in this book, and their potential to illustrate the desertification principles explained in this chapter

Cases	Principles										
	I	II	III	IV	V	VI	VII	VIII	IX	X	XI
Droughts, Aridity and Desertification: The Climatic Factor				**X**		**X**				**X**	**X**
Badlands: What Desertification is Not											
The Harms and Benefits of Dust											
Shrub Encroachment in Drylands: Scientific Basis Regarding Their Positive and Detrimental Effects	**X**		**X**	**X**			**X**				**X**
Wildfires and Desertification: The Role of Fire Regime			**X**			**X**				**X**	
Invasive Plants as a Symptom and Vehicle of Degradation	**X**		**X**	**X**		**X**					**X**
Tree Mortality Processes and Their Role in Accelerating Desertification			**X**	**X**		**X**				**X**	**X**
Moving Livestock to Reverse Desertification Landscapes	**X**		**X**			**X**	**X**			**X**	
The Dehesa: A Cultural Landscape Facing Future Challenges	**X**		**X**	**X**						**X**	**X**
Desertification by Land Abandonment	**X**		**X**	**X**						**X**	**X**
Tourism and Water Stress: A Worrying Convergence in Time and Place	**X**	**X**	**X**		**X**			**X**	**X**		
Inherited Desertification	**X**		**X**	**X**	**X**		**X**			**X**	
Irrigation as a Source of Development and Degradation	**X**	**X**			**X**		**X**	**X**	**X**		
Desertification of the Olive Grove: A Complex Challenge for a Diverse Agricultural System		**X**	**X**	**X**		**X**		**X**	**X**		
Hiding Desertification with Technology: Irrigation and Water Imbalance in Southern Spain	**X**	**X**			**X**		**X**	**X**	**X**	**X**	
The Global Trade of Degradation Distorts the Scope of Desertification		**X**	**X**	**X**	**X**						**X**

food systems to those oriented toward global markets has brought about significant changes in lifestyles, land use, and undesirable environmental and social effects. This transition prioritizes short-term profit maximization as the main driver of decision-making.

This focus on profits fosters fierce competition among producers, who strive to minimize their production costs through various strategies, such as massive production, labor cost reduction, and enhanced marketing. The adoption of new technologies plays a crucial role in these efforts (e.g., developing higher-yield crop varieties, improving irrigation efficiency, and mechanizing labor-intensive tasks). This marks the initial phase of the "Treadmill of Production" (Schnaiberg 1980; Gould et al. 2004), leading to a sequence of events that creates a self-destructive spiral from which escape is difficult.

Implementing these strategies entails increased indebtedness, higher productivity, and greater efficiency. As these innovations spread throughout the production system—because the competitive advantage of early adopters is short-lived and other producers quickly follow suit—the resulting surge in production volume causes product prices to fall. This requires further cost-cutting measures and innovations to generate enough income to cover investments and achieve minimal profits for survival.

This perpetual cycle ensures that only large corporations with substantial financial resources can sustain the pace of innovation, often integrating food production with distribution. Consequently, negative environmental externalities escalate, and employment becomes more precarious. In dryland regions, this dynamic leads to the depletion of aquifers, soil degradation, the hallmarks of desertification processes (Martínez-Valderrama et al. 2023b).

The significant economic development associated with large-scale agricultural production processes (i.e., agro-industry) often leads to episodes of sudden wealth that are widely celebrated by the population (De Stefano and Lopez-Gunn 2012). However, these episodes are relatively ephemeral and can jeopardize the long-term viability of socio-ecosystems. Thus, identifying these bursts of economic growth can serve as an early warning indicator of desertification (Ibáñez et al. 2008a, 2022).

3. *Desertification takes place in coupled land subsystems that are spatially separated.* The socio-economic setup whereby a certain human population thrives by overexploiting its local natural resources has been conventionally called a "hotspot" (e.g., Singh and Ajai 2019). However, this term lumps together causes and consequences and does not allow for explicit functional transferences between desertification subsystems. Such transferences involve matter and energy and can be described using thermodynamics to explain land degradation (del Barrio et al. 2021). We propose to call the subsystem that contains the human population a "foreland" (FL). It is coupled to another land subsystem that contains the natural resource that is being exploited, and that we propose to call "backland" (BL). The coupling is such that the heterotrophic FL self-organizes at the expense of simplifying the autotrophic BL, which performs as an entropy sink and thus becomes degraded. The wider landscape FL and BL are inserted in is not under the target land use, but it is a reservoir of marginal lands where (over)used land is abandoned to and usable land is taken from. We propose to call "hinterland" (HL) to this embedding landscape. These coupled subsystems have own distinct properties and lend well to a system approach describing formal interactions between human populations and their natural environment, such as Social-Ecological Systems (SES) (Ostrom 2009) or Coupled Human And Natural Systems (Alberti et al. 2011). Moreover, their

consideration is crucial to ensure a proper societal awareness. Without it, people often misinterpret the prosperity of the FL.

4. *Desertification has a significant impact on large geographical areas.* A main difference between desertification and other environmental pressures, such as pollution, is that the transferences creating the problem are reversed. For example, in a conventional pollution scheme, the problem is caused by a human setup emitting pollutants to its surroundings at focal locations. However, in desertification the problem is always caused by the surroundings (the BL) becoming exhausted by providing resources to the human setup (the FL). Because of the inherent low efficiency of matter or energy transfers between these subsystems, the size of the BL must be very large with respect to the foreland to keep the overall system working. For example, the case of inherited desertification (Chap. 13) describes the degradation of ca. 80,000 ha during the nineteenthcentury in a mountain range in SE Spain to obtain combustible for lead ore smelting. This BL was one or two orders of magnitude larger than the associated FL made by the exploiting human population at that time. Therefore, when we talk about desertification, we are referring to its effects on a landscape or regional scale extending over hundreds of square kilometers. Isolated local impacts such as the erosion of a hillslope or the pollution created by a farm, critical as they may be, cannot be classified as instances of desertification. Instead, paradigmatic examples of desertification include the desiccation of the Aral Sea, massive soil erosion on the Loess Plateau of China, and the Dust Bowl in the USA. These examples demonstrate the extent and severity of land degradation impact on the environment.
5. *The origin and sink of resources in a desertification setup may be spatially telecoupled.* Environmental degradation in the BL may not necessarily occur on a gradient of proximity to the FL, and may in fact be spatially disconnected. The importance of this aspect increases as the BL production is used for export in a global market scheme rather than for domestic consumption. Illustrative examples include the deforestation of primary forests in South America for the purpose of cultivating soybeans that are subsequently utilized as animal feed to support European livestock (Boerema et al. 2016; Martínez-Valderrama et al. 2021) or the production of fruits and vegetables from southeastern Spain to supply European markets (Martínez-Valderrama et al. 2020b). This lack of spatial contiguity is a crucial aspect of the telecoupling paradigm (Liu et al. 2013), which offers a method to address the spatial disconnection between the primary environmental, socioeconomic, and political factors driving land use change, and the resultant impacts and outcomes of such changes.
   Note that this implies extending the scope of desertification beyond the aridity limits established by the UNCCD definition. This aspect is crucial when designing the LDN, which is implemented at the country level. While degradation and restoration may be balanced within national boundaries, the environmental footprint beyond these borders must also be considered. A good measure of such "LDN-leaks" (Martínez-Valderrama et al. 2025) is the export of degradation, described by the ratio of national land used within a country's territory to

that used outside its borders. On a global scale, this ratio is 73–27%, meaning that nearly three-quarters of what a country consumes is produced domestically, while just over a quarter is imported. However, in some island nations, the balance shifts dramatically. For instance, Japan has a ratio of 8–92%, and the United Kingdom stands at 20–80%. In Europe, the average proportion of land used for food production beyond its borders exceeds 50%.

In terms of BL/FL organization, territories connected over large distances can be represented as a series of regions governed by the spatial scale of the tele-couplings. As illustrated in Chap. 17, the B_1-F_1 system described in point 3 extends to another region, which remains linked to the place of consumption through a second Foreland (F_2).

Consideration of this aspect of desertification is essential, as the trend toward telecouplings is increasing. Over half the world's population currently lives in towns and cities, and by 2050, it is estimated that more than twice as many people will be living in urban areas (6.7 billion) compared to rural settings (3.1 billion) (FAO 2016; UNCCD 2022). This urbanization causes food to travel increasingly greater distances from production sites to consumption locations. Only one-third of the population can fulfil demand with local food crop production (i.e., nearer than 100 km) (Kinnunen et al. 2020), and internationally traded food provides 19% of globally consumed calories (Li et al. 2022).

6. *To detect desertification phenomena in their earliest stages, it is crucial to combine "slow" and "fast" variables of socio-ecosystems.* The former represent the memory of the system and ultimately determine the flows of ecosystem goods and services that people rely on (Walker and Janssen 2002). In contrast, the latter, or rapid environmental variables, are highly sensitive to short-term events, such as rainfall or primary productivity in the pulse-reserve paradigm of drylands (Reynolds et al. 2004), and are thus not useful for characterizing the state of human-environment systems (Reynolds and Stafford Smith 2002). In monitoring of desertification, slow variables help to differentiate signal (trend) from noise (seasonal variability) (Vogt et al. 2011).

   Slow variables, such as the amount of soil organic matter, shape how a fast variable, like crop production, responds to variations in external drivers, such as rainfall during the growing season (Walker et al. 2012). It is the degradation of these slow variables that marks the evolution of desertification since the thresholds of these variables define different steady states of socio-ecological systems (Reynolds et al. 2011). However, rapid changes in land use, driven by fast variables like consumption habits, profitability of land use, or policies influenced by short electoral periods, are the primary cause of land degradation (Friis and Nielsen 2019).

   LDN monitoring, adopted in target 15.3.1 of the SDGs, considers three indicators (Cowie et al. 2018) to cover both slow and fast variables. As a slow indicator of desertification LDN implements monitoring of Soil Organic Carbon (SOC). To detect fast processes, the framework includes land cover and land productivity indicators. Land cover is a highly responsive measure, reflecting land use dynamics, that reveals change in vegetative cover such as through land conversion and resulting habitat fragmentation; land productivity captures relatively fast changes in ecosystem function.

7. *Abrupt changes in land use can foster land degradation in drylands.* Socioeconomic factors, such as positive changes in agricultural value added or agricultural employment, can signal significant transformations in land use (e.g., Puigdefábregas and del Barrio 2000). Drylands have been inhabited regularly for centuries, including land uses like nomadic, farming, oasis, and savannah cultures. However, they are now facing unprecedented pressures from human activities. In those fluctuating environments, sustainability was often achieved by basing exploitation patterns on unfavorable periods, rather than on wet or productive pulses. The inherent resilience of dryland ecosystems to natural variability is threatened by sudden disturbances from land use changes in inappropriate areas, such as the expansion of cultivation on low-production or marginal lands (Lü et al. 2022). This abruptness is increasingly common due to the rapid transformation of territories driven by technological advances and the swift adoption of new lifestyles. For example, the rising demand for superfoods like avocados and quinoa induces rapid land-use changes, which in turn trigger land degradation processes (Jacobsen 2011; Magrach and Sanz 2020; Cho et al. 2021).

   To predict the onset of desertification, it is crucial to consider global phenomena that can rapidly transform a territory, leading to significant environmental consequences. Key drivers such as the influx of investment funds into agriculture (Ouma 2020) and the unequal distribution of income, wealth, and land (Ceddia 2020) must be taken into account. The swift and substantial availability of capital can quickly operationalize agricultural enterprises. This dynamic is a key factor behind the ongoing phenomena of land and water grabbing (Rulli et al. 2013), which generates substantial profits for investors while devastating local socio-economic and environmental systems. For instance, transnational land appropriation exacerbates water insecurity in developing countries, leaving local populations increasingly dependent on food aid and international subsidies, as evidenced in countries like Sudan and Ethiopia (Jackson et al. 2015). It is crucial to recognize this degradation concealed beneath the guise of agricultural and economic development in order to detect desertification in time.

   The transience with which the territory is transformed gives rise to ephemeral episodes of economic splendor that undermines the natural resources on which such enrichment is built. Desertification and development are intrinsically linked, sharing common drivers. It is the management and modulation of these drivers what determine whether a socio-economic system will be sustainable or, on the contrary, is heading toward collapse.

8. *"Fast" socioeconomic variables can lead to a misperception of desertification*, as they may depict a picture of a vigorous and highly productive ecosystem. This is often a consequence of converting the accumulated potential of a mature ecosystem, i.e., its natural capital, into a flow of biological productivity, and subsequently, economic productivity. Ecosystems tend to self-organize to escape from thermodynamic equilibrium, which results in an increase of size,

complexity, and/or information networks (Fath et al. 2004). Opposite to this trend, humans simplify complex ecosystems by favoring usable net primary production over biomass accumulation. This simplification is a form of environmental degradation, which becomes this way associated with any economic exploitation system (del Barrio et al. 2021).

Desertification leads to non-reversible degradation through loss of key ecosystem functions and the depletion of accumulated potential, such as soil or groundwater. Economic profitability may become decoupled from biological productivity when variables such as prices and costs are considered, which can distort the signals of ecosystem exhaustion. While economic enrichment may be high, exploitation can push the socio-ecosystem so far away from its pole of maturity that it cannot return to it, due to the undermining of slow variables. In the steppes of North Africa, we can observe a double example of this transition. The excessive logging of pine trees in Roman times terminated with the possibility of a sparse forest cover, and forced the alpha grass to use asexual reproduction mechanisms to survive (Le Houérou 1995). However, the ecological maturity of these grasslands was so high that they became essential to the region's nomadic livestock farming. The subsequent intensification of livestock farming, supported by animal feed and driven by the high demand of meat, led to the disappearance of large areas of alpha grass, leaving the soil bare and resulting in wind erosion processes that ruined the ecosystem (Martínez-Valderrama et al. 2017).

9. *Aquifers and their associated drainage basins make coupled exploited subsystems of desertification.* At the historical moment when desertification was defined and the UNCCD was established in 1994, cases were restricted to terrestrial surface elements, and groundwater was incidental in their description (for example, the role of bores in settling nomads during the Sahel case). This aligns with the fact that groundwater, despite making up 99% of the Earth's liquid freshwater reserves (Margat and van der Gun 2013), has largely gone unnoticed. Often regarded as an "invisible resource" (United Nations 2022), it has been exploited in a "silent" manner (Llamas and Martínez-Santos 2005). However, groundwater is a slow-renovating natural resource that fits perfectly into a generalized desertification scheme. The extent of irrigated lands is increasing exponentially in drylands everywhere, particularly in emerging or developed economies where such agricultural production is dedicated to exportation, rather than to subsistence or local markets. We observe cases on a global scale where groundwater exploitation has led to prosperous economic periods, resulting in its depletion and degradation (Wada et al. 2010; Famiglietti 2014). Notable examples include arid and hyper-arid regions such as Iran, China, Peru, Spain, North Africa, California, and Saudi Arabia (Martínez-Valderrama et al. 2023b). The connection between groundwater and surface water means that the degradation of the former impacts wetlands, affecting biodiversity in these areas, which are often significant protected zones (Green et al. 2024).

10. *Desertification risk is different from desertification condition.* Desertification manifests through various symptoms of land degradation, including eroded soils, depleted aquifers, loss of vegetation cover, biodiversity decline, and economic impoverishment. In addition to those, further symptoms of social degradation should also be considered, such as famine, unemployment, locked immigration and social unrest. Measuring these variables provides insight into the current condition of a territory. However, condensing these factors into a single indicator has proven, thus far, to be an elusive challenge. Moreover, all those features occur after desertification has triggered, hence their value for early warning is null. Regarding risk, it is essential to identify the drivers that can trigger land degradation. A key factor is changes in land use, one of the components considered in SDG indicator 15.3.1, Proportion of land that is degraded over total land area (Sims et al. 2021). Associated with it are technologies that facilitate access to natural resources, the emergence of markets that incentivize the production of certain products, shifts in lifestyle, and climate change scenarios. Assessing the risk of desertification involves understanding the system's vulnerability and determining the likelihood of a desertification process occurring. This is less straightforward than mapping the current status of desertification, or spotting changes vegetation patchiness that warn of imminent desertification (Kéfi et al. 2007) and presents more challenges, as it requires integrating active desertification processes with potential latent threats. More specifically, it is crucial to understand the mechanisms behind the internal dynamics of socioecosystems, emphasizing the causal, non-linear relationships between latent drivers (such as raw material prices or production costs) that can trigger the deterioration of the resources on which economic activities depend (Ibáñez et al. 2008b). The progression of these processes can be synthesized into a series of early warning indicators that signal the risk of desertification (Ibáñez et al. 2007; Martínez-Valderrama et al. 2016).
11. *One of the main problems facing desertification monitoring and remediation is setting a baseline.* Objective measurement of desertification for large areas has proved extremely difficult, mainly because of multiple criteria and the lack of a reliable baseline (Reeves and Baggett 2014; Cai et al. 2022). Indeed, in order to quantify desertification (its severity and extent), it is necessary to compare the current state of the land with the original, non-degraded state of the ecosystem. As it moves away from that original state it will become more degraded.

The problem with this approach is that it is not always easy to know the original state and that there are degraded states that are functional. For example, the *dehesas* of the southwest of the Iberian Peninsula are degraded versions of the original holm oak and cork oak forests (Ibáñez et al. 2014), and the alpha grass seas of North Africa are the residual version of a pine forest that was felled in Roman times (Le Houérou 1995) and have been functional socio-ecosystems for centuries.

Another issue with this approach is the implicit assumption that any deviation from the original ecosystem constitutes desertification, thereby implying that nothing should be degraded if we aim to completely eliminate desertification. This perspective excludes all forms of human activity, leading to a significant inconsistency (del Barrio et al. 2019). Acknowledging the inevitability of some degree of degradation, the UNCCD proposes, through the LDN framework, to balance the degradation incurred since 2015. In essence, the level of degradation observed in 2015 is deemed acceptable as an arbitrary start point, and serves as the baseline for determining whether further desertification has occurred.

Concluding Remarks

Since the term "desertification" was first introduced nearly a century ago, progress in resolving this serious and pervasive issue has been limited. Despite significant conceptual advancements within the framework of the UNCCD—such as linking its onset to human activities and, more recently, recognizing that the goal should not be to eliminate the degradation inherent to human presence but to attempt to mitigate it—the design of effective solutions continues to be challenged by the complexity and ambiguity of the concept.

With just over 5 years remaining to achieve the Sustainable Development Goals (SDGs), and in light of ongoing biodiversity loss, global warming, and land degradation, there is an urgent need for clear guidelines to direct policy implementation and assist stakeholders in making informed decisions. We believe that the essay presented here is particularly timely in view of recent initiatives such as Land Degradation Neutrality and Convergence of Evidence that require a precise definition of what constitutes desertification.

This work offers a set of guidelines for identifying the complex and often elusive issues associated with desertification. Recognizing that globalization extends these issues beyond drylands, the principles proposed here can also be applied to identify cases of land degradation more broadly. This highlights the need for effective communication to convey the problem accurately and robustly. Desertification is not merely the expansion of deserts or exclusively related to droughts. Despite the persistent use of imagery that continues to add confusion to the issue (Prince and Podwojewski 2019), clarifying this distinction is crucial for designing effective solutions.

In summary, it is essential to accurately identify desertification cases in their early stages to redirect a potentially damaging trend. While this trend may generate short-term benefits, it ultimately condemns a territory to long-term challenges. As climate change remains a significant certainty, maintaining ecosystems in the healthiest possible state is crucial for future resilience.

Acknowledgments This research was financially supported by the project "Plan Complementario de I+D+i en el área de Biodiversidad (PCBIO)" funded by the European Union within the framework of the Recovery, Transformation and Resilience Plan - NextGenerationEU, by the Spanish Ministry of Science, Innovation and Universities and by the Regional Government of Andalucia.

The authors dedicate this work to the memory of Professor Juan Puigdefábregas. The idea for this work arose from the many fruitful and pleasant conversations we had with him.

References

Albaladejo J, Díaz-Pereira E, Vente J De (2021) Eco-holistic soil conservation to support land degradation neutrality and the sustainable development goals. Catena 196:104823. https://doi.org/10.1016/j.catena.2020.104823

Alberti M, Asbjornsen H, Baker LA et al (2011) Research on coupled human and natural systems (CHANS): approach, challenges, and strategies. Bull Ecol Soc Am 92:218–228. https://doi.org/10.1890/0012-9623-92.2.218

Aradóttir AL, Petursdottir T, Halldorsson G et al (2013) Drivers of ecological restoration: lessons from a century of restoration in Iceland. Ecol Soc 18:33. https://doi.org/10.5751/ES-05946-180433

Boerema A, Peeters A, Swolfs S, Vandevenne F (2016) Soybean trade: balancing environmental and socio-economic impacts of an intercontinental market. PLoS One 11:e0155222. https://doi.org/10.1371/journal.pone.0155222

Bond WJ (2019) Open Ecosystems: ecology and evolution beyond the forest edge. Oxford University Press, Oxford

Cai D, Wang X, Jiao L, Geng X (2022) Baseline and status of desertification in Central Asia L Degrad Dev 771–784. https://doi.org/10.1002/ldr.4214

Ceddia MG (2020) The super-rich and cropland expansion via direct investments in agriculture. Nat Sustain 3:312–318. https://doi.org/10.1038/s41893-020-0480-2

Cherlet M, Hutchinson C, Reynolds J et al (2018) World atlas of desertification. Publication Office of the European Union, Luxembourg

Cho K, Goldstein B, Gounaridis D, Newell JP (2021) Where does your guacamole come from? Detecting deforestation associated with the exports of avocados from Mexico to the United States. J Env Manage 278:111482. https://doi.org/10.1016/j.jenvman.2020.111482

Cowie AL, Orr BJ, Castillo Sanchez VM et al (2018) Land in balance: the scientific conceptual framework for Land Degradation Neutrality. Env Sci Policy 79:25–35. https://doi.org/10.1016/j.envsci.2017.10.011

De Stefano L, Lopez-Gunn E (2012) Unauthorized groundwater use: institutional, social and ethical considerations. Water Policy 14:147–160. https://doi.org/10.2166/wp.2012.101

del Barrio G, Sanjuán ME, Hirche A et al (2016) Land degradation states and trends in the Northwestern Maghreb Drylands, 1998–2008. Remote Sens 8:1–25. https://doi.org/10.3390/rs8070603

del Barrio G, Sanjuán ME, Martinez-Valderrama J et al (2019) The paradox of land degradation. In: del Barrio G, Lázaro R (eds) Geoecology and desertification - from physical to human factors. Proceedings of the International Symposium in memory of Prof Juan Puigdefabregas. Estación Experimental Zonas Áridas (CSIC), Almería, Spain, pp 103–105

del Barrio G, Sanjuán ME, Martínez-Valderrama J et al (2021) Land degradation means a loss of management options. J Arid Env 189:104502. https://doi.org/10.1016/j.jaridenv.2021.104502

Famiglietti JS (2014) The global groundwater crisis. Nat Clim Chang 4:945–948. https://doi.org/10.1038/nclimate2425

FAO (2016) Negotiated territorial development in a multi-stakeholder participatory resource planning approach: an initial sustainable framework for the Near East region. Land and Water Division Working Paper 15. FAO, Rome

Fath BD, Jørgensen SE, Patten BC, Straškraba M (2004) Ecosystem growth and development. Biosystems 77:213–228. https://doi.org/10.1016/j.biosystems.2004.06.001

Friis C, Nielsen J (2019) Telecoupling. Exploring land-use change in a globalised world. Palgrave Macmillan, Cham
Gibbs HK, Salmon JM (2015) Mapping the world's degraded lands. Appl Geogr 57:12–21. https://doi.org/10.1016/j.apgeog.2014.11.024
Gould KA, Pellow DN, Schnaiberg A (2004) Interrogating the treadmill of production: everything you wanted to know about the treadmill but were afraid to ask. Organ Env 17:296–316. https://doi.org/10.1177/1086026604268747
Grainger A (2015) Is land degradation neutrality feasible in dry areas? J Arid Env 112:14–24. https://doi.org/10.1016/j.jaridenv.2014.05.014
Green AJ, Guardiola-Albert C, Bravo-Utrera MÁ et al (2024) Groundwater abstraction has caused extensive ecological damage to the Doñana World Heritage Site, Spain. Wetlands 44:20. https://doi.org/10.1007/s13157-023-01769-1
Hill J, Stellmes M, Udelhoven T et al (2008) Mediterranean desertification and land degradation. Mapping related land use change syndromes based on satellite observations. Glob Planet Change 64:146–157. https://doi.org/10.1016/j.gloplacha.2008.10.005
Ibáñez J, Gartzia R, Alcalá FJ, Martínez-Valderrama J (2022) The Importance of prevention in tackling desertification: an approach to anticipate risks of degradation in coastal aquifers. Land 11:1626. https://doi.org/10.3390/land11101626
Ibáñez J, Lavado Contador JF, Schnabel S et al (2014) A model-based integrated assessment of land degradation by water erosion in a valuable Spanish rangeland. Env Model Softw 55:201–213. https://doi.org/10.1016/j.envsoft.2014.01.026
Ibáñez J, Martínez-Valderrama J, Puigdefábregas J (2008) Assessing overexploitation in Mediterranean aquifers using system stability condition analysis. Ecol Model 218:260–266. https://doi.org/10.1016/j.ecolmodel.2008.07.004
Ibáñez J, Martínez-Valderrama J, Puigdefábregas J, Puigdefabregas J (2008) Assessing desertification risk using system stability condition analysis. Ecol Model 213:180–190. https://doi.org/10.1016/j.ecolmodel.2007.11.017
Ibáñez J, Martínez J, Schnabel S (2007) Desertification due to overgrazing in a dynamic commercial livestock-grass-soil system. Ecol Model 205:277–288. https://doi.org/10.1016/j.ecolmodel.2007.02.024
Imeson A (2011) Desertification indicators: from concept to practice. In: Imeson A (ed) Desertification, land degradation and sustainability. Wiley, Hoboken, New Jersey, pp 57–113
Jackson N, Konar M, Hoekstra AY (2015) The water footprint of food aid. Sustainability 7:6435–6456. https://doi.org/10.3390/su7066435
Jacobsen SE (2011) The situation for quinoa and its production in Southern Bolivia: from economic success to environmental disaster. J Agron Crop Sci 197:390–399. https://doi.org/10.1111/j.1439-037X.2011.00475.x
Kéfi S, Rietkerk M, Alados CL et al (2007) Spatial vegetation patterns and imminent desertification in Mediterranean arid ecosystems. Nature 449:213–217. https://doi.org/10.1038/nature06111
Kinnunen P, Guillaume JHA, Taka M et al (2020) Local food crop production can fulfil demand for less than one-third of the population. Nat Food 1:229–237. https://doi.org/10.1038/s43016-020-0060-7
Kosmas C, Kirkby M, Geeson N (1999) Manual on key indicators of desertification and mapping environmentally sensitive areas to desertification. The Medalus project. Mediterranean desertification and land use. European Commission, Luxembourg
Kuderna M, Zsolnay A (2014) Desertification: indicators and thresholds. In: Gliński J, Horabik J, Lipiec J (eds) Encyclopedia of Earth sciences series. Springer, Dordrecht, 200–207 pp
Lavauden L (1927) Les Forêts du Sahara. Rev des Eaux Forêts 65:265–277
Le Houérou HN (1995) Bioclimatologie et biogéographie des steppes arides du Nord de l'Afrique : diversité biologique, développement durable et désertisation.CIHEAM, Montpellier, pp 1–396
Li M, Jia N, Lenzen M et al (2022) Global food-miles account for nearly 20% of total food-systems emissions. Nat Food 3:445–453. https://doi.org/10.1038/s43016-022-00531-w

Li T (2014) What is land? Assembling a resource for global investment. Trans Inst Br Geogr 39. https://doi.org/10.1111/tran.12065
Liu J, Hull V, Batistella M et al (2013) Framing sustainability in a telecoupled world. Ecol Soc 18:26. https://doi.org/10.5751/ES-05873-180226
Llamas MR, Martínez-Santos P (2005) Intensive groundwater use: silent revolution and potential source of social conflicts. J Water Resour Plan Manag 131:337–341. https://doi.org/10.1061/(asce)0733-9496(2005)131:5(337)
Lockeretz W (1978) The lessons of the dust bowl. Am Sci 66:560–569
Lü N, Fu B, Stafford-Smith M et al (2022) Breaking the land degradation-poverty nexus in drylands. Sci Bull 67:2508–2512. https://doi.org/10.1016/j.scib.2022.11.005
Magrach A, Sanz MJ (2020) Environmental and social consequences of the increase in the demand for 'superfoods' world-wide. People Nat 2:267–278. https://doi.org/10.1002/pan3.10085
Margat J, van der Gun J (2013) Groundwater around the world. A geographic synopsis. CRC Press (Taylor & Francis), Boca Raton, FL
Martínez-Valderrama J (2024) Desertificación, cuando el territorio hace aguas [Desertification: lands in troubled waters]. Tirant lo Blanch, Valencia, España
Martínez-Valderrama J, del Barrio G, Sanjuán ME et al (2022) Desertification in Spain: a sound diagnosis without solutions and new scenarios. Land 11:272. https://doi.org/10.3390/land11020272
Martínez-Valderrama J, Gui D, Ahmed Z (2023a) Oasification and desertification under the framework of land degradation neutrality. Env Sci Proc 25:94. https://doi.org/10.3390/ECWS-7-14238
Martínez-Valderrama J, Guirado E, Maestre FT (2020a) Unraveling misunderstandings about desertification: the paradoxical case of the Tabernas-Sorbas Basin in Southeast Spain. Land 9:269. https://doi.org/10.3390/land9080269
Martínez-Valderrama J, Guirado E, Maestre FT (2020b) Discarded food and resource depletion. Nat Food 1:660–662. https://doi.org/10.1038/s43016-020-00186-5
Martínez-Valderrama J, Ibáñez J, del Barrio G et al (2017) Doomed to collapse: why Algerian steppe rangelands are overgrazed and some lessons to help land-use transitions. Sci Total Env 613–614:1489–1497. https://doi.org/10.1016/j.scitotenv.2017.07.058
Martínez-Valderrama J, Ibáñez J, Del Barrio G et al (2016) Present and future of desertification in Spain: implementation of a surveillance system to prevent land degradation. Sci Total Env 563–564:169–178. https://doi.org/10.1016/j.scitotenv.2016.04.065
Martínez-Valderrama J, Olcina J, Gartzia R et al (2023b) Uberizing agriculture in drylands: a few enriched, everyone endangered. Water Resour Manag 38:193–214. https://doi.org/10.1007/s11269-023-03663-1
Martínez-Valderrama J, Parra Valencia JA, Awad T et al (2025) Toward resilient implementation of land degradation neutrality via systemic approaches. Systems 13:408. https://doi.org/10.3390/systems13060408
Martínez-Valderrama J, Sanjuán ME, del Barrio G et al (2021) Mediterranean landscape re-greening at the expense of South American agricultural expansion. Land 10:204. https://doi.org/10.3390/land10020204
Mátyás C, Sun G (2014) Forests in a water limited world under climate change. Env Res Lett 9:085001. https://doi.org/10.1088/1748-9326/9/8/085001
Meyfroidt P, Ryan CM, Aspinall R (2022) Ten facts about land systems for sustainability. PNAS 119:e2109217118
Millenium Ecosystem Assessment (MEA) (2005) Dryland systems. In: Hassan R, Scholes R, Ash N (eds) Ecosystems and human well-being: scenarios. Island Press, Washington, DC., pp 623–662
Needham R (1975) Polythetic classification: convergence and consequences. R Anthropol Inst Gt Br Irel 10:349–369

Orr BJ, Cowie AL, Castillo Sanchez VM et al (2017) Scientific conceptual framework for land degradation neutrality. A report of the science-policy interface. United Nations Convention to Combat Desertification (UNCCD), Bonn

Ortiz EF, Tang G (2005) Review of the Management, Administration and Activities of the Secretariat of the United Nations Convention to Combat Desertification (UNCCD). Geneva

Ostrom E (2009) A general framework for analyzing sustainability of social-ecological systems. Science 325(80):416–419

Ouma S (2020) Farming as financial asset: global finance and the making of institutional landscapes. Agenda Publishing Limited, Newcaste, UK

Prince SD (2016) Where Does Desertification Occur? Mapping Dryland Degradation at Regional to Global Scales. In: Behnke RH, Mortimore M (eds) The End of Desertification? Springer Verlag, Berlin Heidelberg, pp 225–263. https://doi.org/10.1007/978-3-642-16014-1_9

Prince SD, Podwojewski P (2019) Desertification: inappropriate images lead to inappropriate actions. L Degrad Dev 31:677–682. https://doi.org/10.1002/ldr.3436

Puigdefabregas J (1995) Desertification: stress beyond resilience, exploring a unifying process structure. Ambio 24:311–313

Puigdefábregas J, del Barrio G (2000) SURMODES: a surveillance system for assessing and monitoring of desertification. Expo Hannover 2000 World Project. http://www.eeza.csic.es/surmodes/surmodes/welcome.htm

Puigdefabregas J, Barrio G, Hill J (2009) Ecosystemic approaches to land degradation. In: Romero-Díaz A (ed) Advances in studies on desertification. Servicio Publicaciones Universidad de Murcia EDITUM, Murcia, pp 77–88

Reeves MC, Baggett LS (2014) A remote sensing protocol for identifying rangelands with degraded productive capacity. Ecol Indic 43:172–182. https://doi.org/10.1016/j.ecolind.2014.02.009

Reynolds JF (2021) Desertification is a prisoner of history: an essay on why young scientists should care. Ecosistemas 30:2302. https://doi.org/10.7818/ECOS.2302

Reynolds JF, Grainger A, Smith DMS et al (2011) Scientific concepts for an integrated analysis of desertification. L Degrad Dev 183:166–183

Reynolds JF, Kemp PR, Ogle K, Fernández RJ (2004) Modifying the "pulse-reserve" paradigm for deserts of North America: precipitation pulses, soil water, and plant responses. Oecologia 141:194–210. https://doi.org/10.1007/s00442-004-1524-4

Reynolds JF, Stafford Smith M (2002) Do humans cause deserts? In: Reynolds J, Stafford Smith M (eds) Global desertification. Do human cause deserts. Dahlem University Press, pp 1–21

Rulli MC, Saviori A, Odorico PD (2013) Global land and water grabbing. PNAS 110:892–897. https://doi.org/10.1073/pnas.1213163110

Safriel U (2017) Land degradation neutrality (LDN) in drylands and beyond – where has it come from and where does it go. Silva Fenn 51:1–19. https://doi.org/10.14214/sf.1650

Sanjuán ME, Martínez-Valderrama J, Ruiz A, del Barrio G (2022) Land use intensification affects the relative importance of climate variation and active land degradation: convergence of six regions around the world. L Degrad Dev, https://doi.org/10.1002/ldr.4326

Schnaiberg A (1980) The environment: from surplus to scarcity. Oxford University Press, New York

Schubert SD, Suarez MJ, Pegion PJ et al (2004) On the cause of the 1930s dust bowl. Science 303:1855–1859. https://doi.org/10.1126/science.1095048

Sims NC, Newnham GJ, England JR et al (2021) Good practice guidance. SDG Indicator 15.3.1, proportion of land that is degraded over total land area. Version 2.0. United Nations Convention to Combat Desertification, Bonn

Singh RB, Ajai (2019) A composite method to identify desertification 'hotspots' and 'brightspots'. L Degrad Dev 30:1025–1039. https://doi.org/10.1002/ldr.3290

Turner MD, Davis DK, Yeh ET et al (2023) Great green walls: hype, myth, and science. Annu Rev Env Resour 48:263–287. https://doi.org/10.1146/annurev-environ-112321-111102

United Nations (2022) The United Nations World Water Development Report 2022: Groundwater: Making the invisible visible. UNESCO, Paris

UNCCD (2016) Land degradation neutrality: the target setting programme. UNCCD, Bonn
UNCCD (2022) Global land outlook. Land restoration for recovery and resilience, 2nd ed. UNCCD, Bonn
UNCCD (United Nations Convention to Combat Desertification) (1994) United Nations Convention to Combat Desertification in those countries experiencing serious drought and/or desertification, particularly in Africa. UNCCD, Paris
Vogt JV, Safriel U, Von Maltitz G et al (2011) Monitoring and assessment of land degradation and desertification: towards new conceptual and integrated approaches. L Degrad Dev 22:150–165. https://doi.org/10.1002/ldr.1075
Wada Y, Van Beek LPH, Van Kempen CM et al (2010) Global depletion of groundwater resources. Geophys Res Lett 37:L20402. https://doi.org/10.1029/2010GL044571
Walker BH, Carpenter SR, Rockstrom J et al (2012) Drivers, slow variables, fast variables, shocks, and resilience. Ecol Soc 17:1–4. https://doi.org/10.5751/ES-05063-170330
Walker BH, Janssen MA (2002) Rangelands, pastoralists and governments: interlinked systems of eople and nature. Philos Trans R Soc Biol Sci 357:719–725. https://doi.org/10.1098/rstb.2001.0984
Wang T, Wu W, Xue X et al (2004) Study of spatial distribution of sandy desertification in North China in recent 10 years. Sci China Ser D Earth Sci 47:78–88. https://doi.org/10.1360/04zd0009
Wessels KJ, Prince SD, Carroll M, Malherbe J (2007) Relevance of rangeland degradation in semiarid Northeastern South Africa to the nonequilibrium theory. Ecol Appl 17:815–827. https://doi.org/10.1890/06-1109

Chapter 2
Droughts, Aridity and Desertification: The Role of Climate Variations

Jorge Olcina Cantos and Javier Martí Talavera

Abstract Desertification and droughts are complex processes that integrate physical and human aspects, erroneously presented as coeval in the genesis of landscapes that structurally have no vegetation cover, degraded soils and hardly any humidity. The circumstantial situation (drought sequence) is confused with the geographical structural condition (aridity) as processes that favour the loss of vegetation and soil vitality in a geographic area. In Spain, due to climatic conditions, different types of droughts occur, related to a diverse evolution of regional atmospheric dynamics. The south-east of the Iberian Peninsula is the geographical area most frequently affected by drought periods, due to the confluence of atmospheric factors and its geographic location. In addition, geological, lithological and geomorphological factors also determine a state of natural aridity. However, aridity is not synonymous with desertification. Desertification requires the intervention of human factors and its territorial manifestation can exceed the geographic boundaries of arid areas. The future climate development in many Spanish regions suggests a scenario of lower rainfall and greater irregularity in its genesis, higher actual evaporation and potential evapotranspiration rates and greater water stress; all of which favour the spread of steppe and desert climates and, if human activity does not adapt to this new climate reality, accelerated desertification processes.

Keywords Droughts · Desertification · Soil erosion · Aridity · Climate conditions · Future conditions

J. Olcina Cantos (✉)
Department of Regional Geographic Analysis and Physical Geography, Universidad de Alicante, Alicante, Spain
e-mail: jorge.olcina@ua.es

J. Martí Talavera
Instituto Multidisciplinar para el Estudio del Medio "Ramón Margalef", Universidad de Alicante, Alicante, Spain
e-mail: javier.marti@ua.es

J. Martínez-Valderrama and J. Olcina Cantos (eds.), *The Labyrinth of Desertification*,
https://doi.org/10.1007/978-3-032-09648-7_2

Droughts as Part of a Cocktail Called Desertification

We have become accustomed to seeing images in the media and publications of cracked earth at the bottom of reservoirs with hardly any water during drought periods with headlines claiming that Spain 'is becoming a desert' (Tello 2021) (Fig. 2.1). There is an incorrect association between drought and desertification, which are understood as simultaneous processes that have the ultimate effect of transforming a geographical space into a territory without vegetation, with degraded soil, no humidity and with a continuous process of biogeographical deterioration. The circumstantial (drought) and structural (aridity) states that shape or model a landscape are confused (Beguería et al. 2025).

If an arid territory suffers the continued effects of a lack of precipitation over a long period of time (decades), the ground can experience a degradation of its physical and chemical components until a period of no return is reached, culminating in a process of desertification due, exclusively, to natural causes. The same is the case of the aquifers existing within the territory. To this, we should add the human factors that aggravate this natural process. There are two important premises that should be fulfilled in the latter state: the condition of aridity and events of prolonged drought over a long period of time. In short, this indicates a change in the climate conditions of the territory (Carvalho et al. 2022). Therefore, aridity is not a synonym of desertification. Desertification requires the intervention of the human factor. In fact, the desertification processes can be triggered when attempts are made to 'combat' aridity or to resolve drought through, for example, intensive changes in land use (from rain-fed to irrigated) on a large geographical scale. Hence, development and desertification come to share the same driving agents. If development ignores the signs of environmental deterioration, a no-return desertification process could begin (Fig. 2.2).

Fig. 2.1 Cracked earth in the Amadorio Reservoir in 2014 (Villajoyosa, Alicante). Chronic association between drought, aridity and desertification. Photo by Jorge Olcina

Fig. 2.2 Loam gullies in Pyrenean foothills (Araguás del Solano, Huesca). Source: Taken from García Ruiz and Lasanta (2018)

Aridity, a Geographical Characteristic of the Territory

The condition of aridity in a geographical space does not depend solely on climate features. It is true that the relationship between high temperatures that favour the evaporation of atmospheric humidity and scarce rainfall favours a higher degree of aridity in a territory. This means that aridity is a condition acquired by territories over long periods of time (centuries) with certain climate conditions. However, aridity is also related to the geomorphological features of a territory (lithology, forms, slope). The lithological composition of a geographic space is fundamental for establishing vegetation and shaping a natural landscape. The fine-fraction sedimentary lithologies (clays, loams, sands) hinder the development of mature soils and give rise to landscapes with sparse vegetation cover. However, this does not mean that these landscapes do not have biogeographical wealth and a high level of biodiversity adapted to the features of the soil where they are located (Paniagua et al. 2019).

If soft lithologies converge with climate conditions of scarce rainfall and high evaporation, the degree of aridity of a territory increases. This occurs in wide areas of south-east Spain, but also in areas of the central sector of the Ebro valley or the depressions of the Duero, Tajo and Guadiana. Furthermore, if these lithologies are manifested in the form reliefs with a steep slope, these spaces are usually affected by erosive processes, preventing the formation of mature soils. Therefore, there are

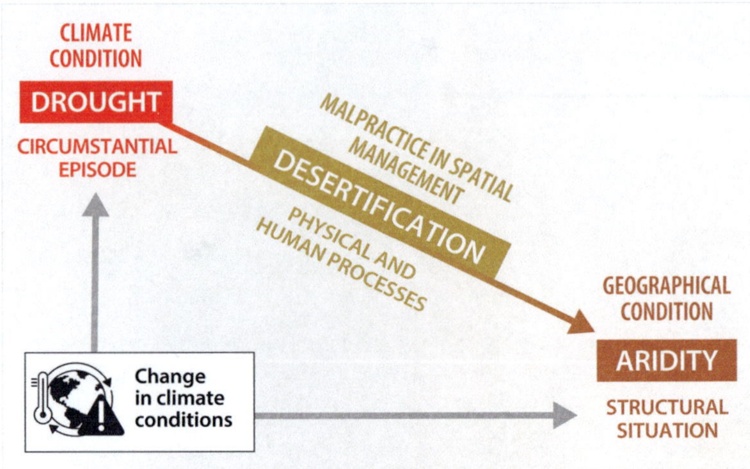

Fig. 2.3 Drought, aridity, desertification. Source: Own elaboration

territories in mountainous areas with abundant average rainfall but with an arid physiognomy due to the runoff in areas with steep slopes (Fig. 2.2).

In the Canary Islands, the volcanic lithologies and scarce rainfall recorded in the sectors not exposed to trade winds (horizontal precipitation related to the sea of clouds) favour the development of arid landscapes, particularly in the central (Tenerife, Gran Canaria) and eastern islands. In addition, the volcanic soils are transformed in areas of commercial agriculture that exploit the aquifer resources in the volcanic reliefs; a process that would increase the future possibility of generating areas of desertification if the agricultural activity were to be abandoned (García-Rodríguez et al. 2016) (Fig. 2.3).

Drought or Droughts in Spain?

The association between droughts and desertification is not, therefore, correct. These processes have a different duration, the physical state of the territories in which they occur is different (circumstantial and structural) and they have causal agents that are diverse and not directly atmospheric. First, in Spain, we cannot refer to 'drought' in the singular. There are different types of drought depending on the territory affected and the atmospheric causes leading to these sequences of lower rainfall with respect to the average annual values of each of them. In 2001, Olcina conducted an initial typology of drought in Spain after analysing the effects of the dry sequence, with different regional characteristics, occurring in the country between 1989 and 1995. Thus, three types of droughts were identified in Spain with different effects on the availability of water resources and the socio-economic impact: Iberian droughts, Cantabrian droughts and south-eastern droughts, to which the drought sequences occurring in the Canary archipelago were

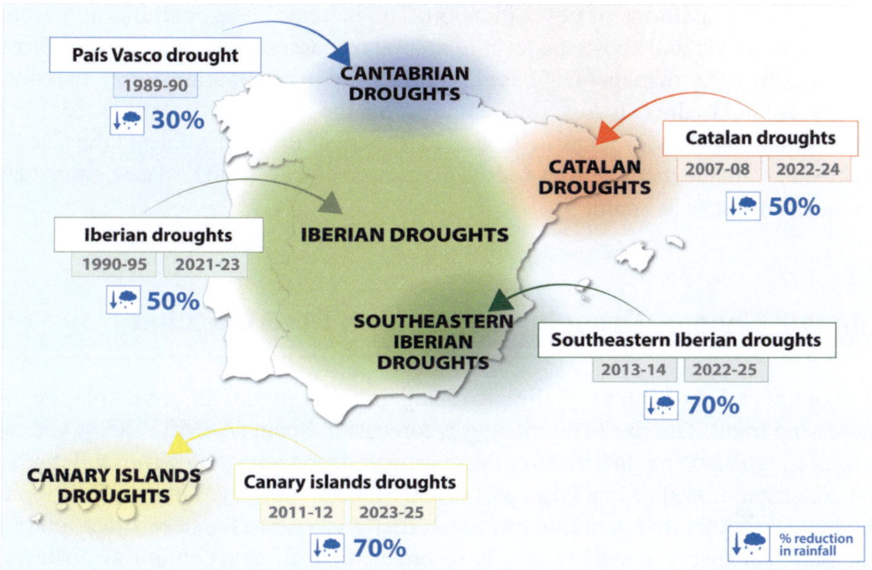

Fig. 2.4 Typology of droughts in Spain. Own elaboration

added, which have a chronological similarity with the south-eastern droughts (Olcina 2001). To these types of drought we can add the sequences affecting the north-east of the peninsula (Catalonia) which also have a specific atmospheric cause (Fig. 2.4).

In general, we can refer to a dry sequence when the presence of an anticyclone in the atmosphere over the Iberian peninsula exceeds 150 days and 50% of these days occur between March and April and between October and December. We can also add the development of atmospheric configurations prone to a scarce rainfall on a regional level. For example, the entrance of Atlantic storms that drop rain on a large part of the peninsula (west and centre) has very little rainfall effectiveness on the Mediterranean coast, particularly in the south-east of the Iberian peninsula. Hence, this atmospheric configuration, which is frequent in the cold months of the year in Spain, favours the persistence of drought conditions of the south-east of the peninsula (Trullenque-Blanco et al. 2024).

Studies have been carried out on the relationship between the drought sequences occurring in the Spanish territory, with atmospheric oscillation mechanisms occurring in the Pacific basin (ENSO), the Atlantic (NAO) or the Mediterranean (WeMo). The Iberian droughts are linked to the negative phase of the Southern Oscillation Index, the Cantabrian droughts are associated with a positive NAO index (Olcina 2001) and the droughts in the Mediterranean regions are related to the positive Mediterranean Oscillation (Oliva et al. 2006). However, these associations are not conclusive due to the geographical characteristics of the Spanish territory.

From the second half of the twentieth century to the present day, we can say that the development of drought atmospheric sequences have not generated conditions of desertification in Spain by themselves. Human factors have had a greater weight in generating conditions of desertification. The increase in agricultural and urban consumption over and above the level of natural resources existing in the territories, the deficient water management based on principles of continuous supply that does not correspond to the existing climate reality and the loss of fertile soil in the urban transformations of many areas in the east and south of the peninsula and the Canary archipelago are the main reasons for the generation of 'desertified' spaces, more than drought sequences in Spain.

Climate Change, Droughts, Aridity and Desertification

Within the current context of climate change, aridity processes are displaying an increasing trend. The rise in temperatures forecast in Spain (AEMET 2024) and the greater irregularity in rainfall with a decreasing trend in many river basins will favour the condition of aridity in a large part of the Spanish territory. Thus, in addition to the human factors that generate processes that give rise to the desertification of a geographical space, now there are also more adverse climate conditions, fostering intense droughts and aridity processes. The result will be the creation of larger areas of potential desertification in Spain.

In Iberian latitudes, climate change is not only manifested in an increase in temperatures and an overall loss of thermal comfort but also in a more irregular behaviour of precipitations and the alteration of seasonal rainfall patterns (Cramer et al. 2018; Meseguer-Ruiz and Olcina 2023). In addition, extreme weather events related to temperature and precipitation are becoming more intense and frequent. The latter is related to the intensification of the global energy readjustment processes within the context of a maladjusted energy balance and with the warming of the seas that surround the mainland territory, particularly the Mediterranean Sea basin (Olcina et al. 2024).

The temperature forecasts for moderate or high emissions scenarios indicate an increase in the average annual temperature and the number of hot days per year in a large part of the Spanish territory. The Cantabrian regions would experience a less abrupt change than the rest of the country (Adaptecca 2025). The forecast increase in summer heat and the duration of heatwaves would foster an increase in the conditions of real evaporation and evapotranspiration. The latter have values of over 110 mm/day per year in a larger part of the south of the peninsula (Tajo, Guadiana, Guadalquivir basins and the south-east of the peninsula). In short, the thermal evolution forecast for the country would lead to a reduction in groundwater reserves (Fig. 2.5).

Different studies have analysed the change in the seasonal rainfall patterns in Spain (Ruiz-Sinoga et al. 2011; Acero 2012; Serrano-Notivoli 2017; CEDEX 2021; Senent-Aparicio et al. 2023), highlighting the 'Mediterraneanisation' process of the maximum precipitations towards the west of the peninsula (Cordillera Ibérica sector) with a main peak in autumn (González-Hidalgo et al. 2010).

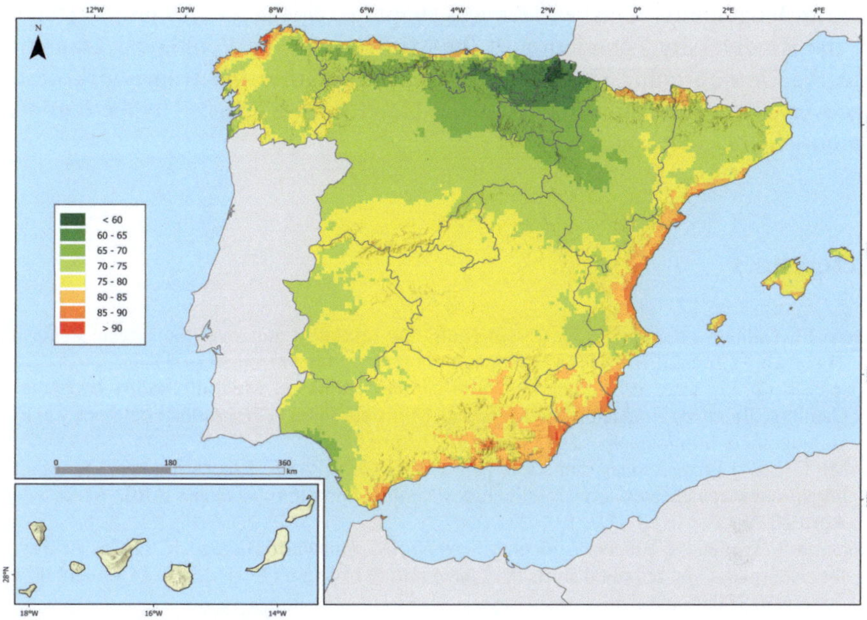

Fig. 2.5 Forecast of the number of hot days in mainland Spain and the Balearic Islands (RCP 8.5). Horizon 2041–2070. Abnormality with respect to the base period (1971–2000). (*Hot days: Number of days in a period time whose maximum temperature is above the 90th percentile of a climate period of reference) Source: Visor AdapteCCa. Available at: www. http://adaptecca.es/

Drought has become an important study topic for water planning in Spain (Vicente-Serrano 2021). There is uncertainty as to the duration of droughts which have been occurring since the beginning of the current century. Contrary to the long sequences that occurred in the second half of the twentieth century that generated droughts in the Iberian Peninsula (Iberian droughts), with a duration of between 3 and 5 years (calendar year), the droughts occurring in the twenty-first century are more intense, but with a shorter duration (1 or 2 calendar years) (González et al. 2020; Torelló-Sentelles et al. 2022; Trullenque-Blanco et al. 2024).

In addition to the natural effects of climate change, such as a lower availability of water, we can add the human activities that intensify the water stress in Spain, such as the increase in agricultural and urban use. The current water plans (2021–2027) have a traditional focus based on extending the supply of water resources, even though the MITERD recommends reducing the use of water by 15% before 2050. This context will give rise to a scenario of greater competition for a resource that is becoming increasingly scarcer, with more frequent extreme hydrological episodes and more arid territories. Human decisions will be fundamental for reducing the impacts of climate change and preventing critical reversibility thresholds.

Acknowledgements This research was supported by the ATLAS project (funded by the Biodiversity Foundation of the Ministry for the Ecological Transition and the Demographic Challenge (MITECO) within the framework of the Recovery, Transformation and Resilience Plan (PRTR), financed by the European Union—NextGenerationEU).

References

Acero FJ, Gallego MC, García JA (2012) Multi-day rainfall trends over the Iberian Peninsula. Theor Appl Climatol 108:411–423. https://doi.org/10.1007/s00704-011-0534-5

Adaptecca (2025) Visor de Escenarios sobre cambio climático. Madrid, Oficina Española de Cambio Climático, Fundación Biodiversidad. Ministerio para la Transición Ecológica y el Reto Demográfico. http://adaptecca.es. Accessed 22 April 2025

AEMET (2024) Proyecciones climáticas para el siglo XXI. AR6: Escenarios PNACC. Madrid. https://www.aemet.es/es/serviciosclimaticos/cambio_climat/proyecciones-AR6. Accessed 22 April 2025

Beguería S, Trullenque-Blanco V, Vicente-Serrano SM, González-Hidalgo JC (2025) Aridity on the rise: spatial and temporal shifts in climate aridity in Spain (1961–2020). J Climatol https://doi.org/10.1002/joc.8775

Carvalho D, Pereira S, Silva R, Rocha A (2022) Aridity and desertification in the Mediterranean under EURO-CORDEX future climate change scenarios. Clim Change 174. https://doi.org/10.1007/s10584-022-03454-4

CEDEX (2021) Impacto del cambio climático en las precipitaciones máximas en España, Madrid. Centro de Estudios Hidrográficos, 404 pp

Cramer W, Guiot J, Fader M et al (2018) Climate change and interconnected risks to sustainable development in the Mediterranean. Nat Clim Change 8:972–980. https://doi.org/10.1038/s41558-018-0299-2

García-Rodríguez JL, García-Rodríguez FJ, Castilla-Gutiérrez C (2016) Human heritage and sustainable development on arid islands: the case of the eastern Canary islands. Isl Stud J 11:113–130. https://doi.org/10.24043/isj.337

García Ruiz JM, Lasanta T (2018) El Pirineo Aragonés como paisaje cultural. Pirineos, 173:e038. https://doi.org/10.3989/pirineos.2018.173005

González-Hidalgo JC, Brunetti M, de Luis M (2010) Precipitation trends in Spanish hydrological divisions, 1946–2005. Clim Res 43:215–228. https://doi.org/10.3354/cr00937

González A, Santos M, Hernández JC, Acosta B, Montalvo J (2020) Aumenta la frecuencia e intensidad de las sequías por el cambio climático en España. Fundación Matrix. Proyecto Climvac. https://fundacionmatrix.es/aumenta-la-frecuencia-e-intensidad-de-la-sequia-por-el-cambio-climatico-en-espana/. Accessed 22 April 2025

Meseguer-Ruiz O, Olcina Cantos J (2023) Climate change in two Mediterranean climate areas (Spain and Chile): evidences and projections. Investig Geogr 79:9–31

Olcina Cantos J (2001) Tipología de sequías en España Ería nº 56, p 201–227

Olcina Cantos J, Martí Talavera J, Sánchez Almodovar E (2024) Evol reciente de precipitación y temperatura en la región mediterránea. Cuadernos Geográficos 63(2):51–73

Oliva M, Lopez-Bustins J, Barriendos M, Muedra C, Martin-Vide J (2006) Reconstrucción histórica de la Oscilación del Mediterráneo Occidental (WeMO) e inundaciones en el levante peninsular (1500–2000). In: Cuadrat Prats, JM, Saz, MA, Vicente Serrano S, Lanjeri S de Luis Arrillaga M, González Hidalgo JC (eds) En Clima, Sociedad y Medio Ambiente. Publicaciones de la Asociación Española de Climatología (AEC), p 1–11

Paniagua L, García-Martín A, Moral F et al (2019) Aridity in the Iberian Peninsula (1960–2017): distribution, tendencies, and changes. Theor Appl Climatol 138:811–830. https://doi.org/10.1007/s00704-019-02866-0

Ruiz-Sinoga J, García-Marín R, Martínez Murillo J, Gabarrón-Galeote M (2011) Precipitation dynamics in southern Spain: trends and cycles. Int J Climatol 31:2281–2289. https://doi.org/10.1002/joc.2235

Senent-Aparicio J, López-Ballesteros A, Jimeno-Sáez P, Pérez-Sánchez J (2023) Recent precipitation trends in Peninsular Spain and implications for water infrastructure design. J Hydrol 45:101308. https://doi.org/10.1016/j.ejrh.2022.101308

Serrano Notivoli R (2017) Reconstrucción climática instrumental de la precipitación diaria en España: Ensayo metodológico y aplicaciones. Tesis doctoral, Universidad de Zaragoza

Tello B (2021) España se seca, el desierto avanza. Madrid. Ed. Universidad de Mayores Experiencia Recíproca, p 22

Torelló-Sentelles H, Franzke CLE (2022) Drought impact links to meteorological drought indicators and predictability in Spain. Hydrol Earth Syst Sci 26:1821–1844. https://doi.org/10.5194/hess-26-1821-2022

Trullenque-Blanco V, Beguería S, Vicente-Serrano SM et al (2024) Catalogue of drought events in peninsular Spanish along 1916–2020 period. Sci Data 11:703. https://doi.org/10.1038/s41597-024-03484-w

Vicente-Serrano S (2021) The evolution of climatic drought studies in Spain over the last few decades. Geographicalia 7–34. https://doi.org/10.26754/ojs_geoph/geoph.2021734640

Chapter 3
Badlands, What Is Not Desertification

Estela Nadal-Romero and Adolfo Calvo-Cases

Abstract Badlands are among the most spectacular erosive landforms in the world, typically associated with semi-arid regions but also present in more humid environments. Although often mistaken for degraded landscapes, they are primarily the result of natural geomorphological and hydrological processes acting over geological timescales. Despite their visual harshness and high erosion rates, they do not fall within the definition of desertification as established by the United Nations Convention to Combat Desertification (UNCCD), since their formation is not predominantly driven by human activity. This paper provides an overview of the main types of badlands, their distribution across the Iberian Peninsula and beyond, and the geomorphic, hydrological, and ecological processes that define them. Particular attention is paid to their sediment dynamics, ecological functions, and potential role as indicators of environmental change. Disentangling natural erosive processes from anthropogenic impacts is key to advancing our understanding of landscape dynamics in sensitive environments

Keywords Badlands · Geomorphology · Mediterranean · Soil erosion · Semi-arid environments · Sediment dynamics

E. Nadal-Romero (✉)
Instituto Pirenaico de Ecología (IPE-CSIC), CSIC, Zaragoza, Spain
e-mail: estelanr@ipe.csic.es

A. Calvo-Cases
Inter-University Institute for Local Development, Universitat de València, Valencia, Spain
e-mail: adolfo.calvo@uv.es

J. Martínez-Valderrama and J. Olcina Cantos (eds.), *The Labyrinth of Desertification*,
https://doi.org/10.1007/978-3-032-09648-7_3

Badlands as Natural Geomorphological Features

There are geological landforms that have nothing to do with desertification. These include phenomena such as badlands, which are particularly common in Spain. According to the Encyclopaedia of Geomorphology, "badlands are deeply dissected erosive landscapes formed in soft rocks, generally in semi-arid regions, although not exclusively" (Harvey 2004, p. 45). Badlands are natural erosional landforms shaped primarily by geomorphological and climatic processes over long geological timescales. They are characterized by a succession of gullies or ravines, often with sharp ridges in rugged terrain and sparse vegetation cover. Despite their hostile and harsh appearance, badlands are not considered an example of desertification, as they are natural erosional features rather than a result of human-induced land degradation.

In general, badlands are associated with semi-arid environments, but they can also develop in wetter climates. Typically, three types of badlands are identified based on climatic conditions (Gallart et al. 2002): (i) arid badlands, which are found in areas where annual precipitation is less than 200 mm (not present in Spain, but prevalent in North African and Middle Eastern areas). These badlands are considered "physical" badlands, as the arid conditions prevent effective vegetation control, and geomorphic processes are entirely governed by the bedrock; (ii) semi-arid badlands, which receive between 200 and 700 mm of annual rainfall, often with origins dating back several thousand years; and (iii) humid badlands, which receive more than 700 mm of rainfall and have extremely high dynamics. Due to geological setting and climatic conditions, the south-eastern part of the Iberian Peninsula contains the best and most extensive examples of these geoforms: they are located in the Guadix and Baza depressions (Granada), the Fortuna and Mula basins (Murcia), the Tabernas Desert (Figs. 3.1 and 3.2), the Vera and the Almanzora river basins (Almería), and in the Ebro Depression in Bardenas (Navarra) (Fig. 3.3) and Monegros (Huesca). They are also found, albeit to a lesser extent, in the marl outcrops of the Aragonese and Catalan Pyrenees, between Pamplona, Jaca, and Sabiñánigo, as well as in Aínsa, Campo (Huesca), the Tremp Basin (Lleida), and Vallcebre (Girona).

Significant badland formations can also be found in other parts of Europe. Their global distribution includes notable examples in southern Italy (calanchi) (Phillips 1998) and southern France (Mathys et al. 1996), representing variations of badland landscapes in different lithological and climatic settings. Outside Europe, striking examples include the United States (such as the Badlands National Park in South Dakota) (Stetler et al. 2011) and China (Zhangye Danxia Landform, known for its vivid colors and stratified erosion) (Peng 2001).

Fig. 3.1 Panoramic view of the El Cautivo basin, Tabernas Desert (Almería), covering 350 km². From the elevated terrain on the right, a series of paleo-pediments extends across the landscape. The most widespread surface dates back approximately 14,000 years, and the erosional incision has occurred in several phases during this period, following the downcutting of the Tabernas dry river (rambla), triggered by a series of tectonic steps and aided by the steep altitudinal gradient of the basin (Alexander et al. 2008). Photo by Adolfo Calvo-Cases

Fig. 3.2 Incision of the Tabernas dry river (rambla) into the Tortonian marls that fill the basin of the same name. The yellowish, flat-surfaced levels on the right bank correspond to lacustrine deposits, the upper surface of which has been dated to 14,000 years ago and formed following the tectonic stepping of the basin. Photo by Adolfo Calvo-Cases

Fig. 3.3 Located between Navarra and Aragón, the Bardenas Reales (418 km²) form the largest area of badlands in the Iberian Peninsula. Erosive activity is linked to the Quaternary incision phases of the Aragón and Ebro rivers (Desir and Marín 2013). Photo by Estela Nadal-Romero

Geomorphological, Hydrological, and Ecological Dynamics in Badlands

In general, badlands are not the result of human activity and degradation. In south-eastern Spain, their formation is linked to a reactivation of erosion associated with a relatively recent tectonic uplift, with periodic pulses (Díaz-Hernández and Juliá 2006). In some cases, however, other, more secondary factors, such as human activity (i.e., historical deforestation, overgrazing) or abrupt climatic changes, cannot be completely ruled out. This is the case of the badlands located in the Pyrenean or Central System, where tectonic stability exists, and erosive reactivation must be explained by other factors. However, even in these cases, the degree of human influence does not reach the thresholds typically associated with desertification as defined by the United Nations Convention to Combat Desertification (UNCCD 1994).

With few exceptions, badlands form on soft or easily erodible rocks, characterized by steep slopes that promote rapid landscape dissection and severe erosion. They are associated with unconsolidated, poorly cemented lithologies, with clay and marl outcrops being the most representative. The hydrological behavior of these systems is dominated by Hortonian overland flow and concentrated runoff, which accelerate rill and gully formation (Cantón et al. 2001). The role of subsurface runoff in developing piping, on salt rich materials also contributes to the development of gullies and badland landforms (see Faulkner et al. 2004). From a hydrological and geomorphological perspective, badlands are highly dynamic and significant sediment producers, being considered sediment production hotspots (Nadal-Romero et al. 2011). The high sediment availability has implications not only for landscape evolution but also for downstream hydrological systems, including dam siltation and floodplain dynamics (Syvitski and Kettner 2011). Their landscapes are characterized by an abundance of short-length ravines, steep slopes, a lack of vegetation, and an overall impression of extreme terrain, despite the fact that rainfall is sufficient to support dense arboreal vegetation.

However, from an ecological point of view, badlands are important because they support plant and animal species adapted to these extreme conditions (Martínez-Paz et al. 2022). The seed bank in badlands is rich at the surface, consisting of small erosion-adapted seeds that become trapped in the numerous cracks and roughness of certain landforms (García-Fayos and Recatalá 1992). This may enhance their survival by reducing the risk of being washed away by runoff. However, due to the inherent characteristics of badlands, their extreme hydro-geomorphological dynamics, and limited water availability, vegetation struggles to establish successfully on the steeper slopes. Additionally, the presence of biological soil crusts, composed of cyanobacteria, lichens, and mosses, plays a key role in stabilizing the soil surface, enhancing moisture retention, and supporting early successional stages (Lázaro et al. 2022). In fact, some badlands are now integrated into conservation frameworks such as Natura 2000 due to their unique habitats.

Conservation, Scientific Value, and Perspectives on Landscape Management

Badlands are natural geoforms shaped by erosive processes over time. The landscapes created by intense erosion and the rapid evolution of slopes and channels are truly spectacular. They are characterized by steep slopes with nearly vertical walls, ruiniform terrains, and intricate microforms on the slopes, including micropinnacles and small bulges, known as popcorn, sometimes displaying a striking diversity of colors. Because of their beauty and uniqueness, many badlands have become popular tourist attractions and, in some cases, have been designated as protected zones where any form of intervention is strictly prohibited. Their aesthetic appeal and scientific significance have led to their inclusion in geotourism initiatives and geoparks (Zgłobicki et al. 2019). For example, the Tabernas Desert is not only a geomorphological reference point but also a protected area under the European Landscape Convention.

In some contexts, the possibility of restoring these landscapes is discussed, but such interventions would be costly and largely ineffective. Thus, the cost, scale, and technical limitations of restoration in badlands render most interventions impractical (Valentin et al. 2005). Despite their high erosion rates and significant sediment production, restoration efforts should be focused only on specific areas, primarily to limit connectivity between slopes and channels (e.g., via check dams). Rather than attempting to restore them, it is suggested that badlands should be allowed to develop naturally, serving as natural laboratories for various scientific disciplines (where hydrological and geomorphological processes can be observed and studied in real time) and as excellent sites for environmental education. In the context of global change, badlands offer exceptional opportunities to study the interactions between rainfall, runoff, and erosion. They also allow us to appreciate the beauty of a wild landscape and provide valuable insights into the hydrological and geomorphological processes that shape our planet.

Badlands are striking natural features that should be appreciated not as symptoms of degradation, but as evidence of the Earth's dynamic processes. Looking ahead, badlands may serve as key indicators in the monitoring of environmental change. Their high sensitivity to shifts in rainfall patterns, temperature, and land cover makes them valuable natural laboratories for understanding the effects of climate change on soil erosion and landscape dynamics (Nadal-Romero et al. 2021). Long-term monitoring of badland systems can inform broader discussions on land degradation neutrality (LDN) and contribute to the design of more nuanced environmental policies that distinguish between natural and anthropogenic major ways of land degradation.

Acknowledgments This research was supported by the "Geoenvironmental Processes and Global Change" (E02_23R) research group is financed by the Aragón Government and the European Social Fund (ESF-FSE). The authors wish to thank José María García Ruiz for his valuable review and insightful comments on this chapter.

References

Alexander RW, Calvo-Cases A, Arnau-Rosalén E, Mather AE, Lázaro-Suau R (2008) Erosion and stabilisation sequences in relation to base level changes in the El Cautivo badlands, SE Spain. Geomorphology 100(1–2):83–90. https://doi.org/10.1016/j.geomorph.2007.10.025

Cantón Y, Solé-Benet A, Domingo F (2001) Temporal and spatial patterns of soil water availability in semiarid badlands of SE Spain. Catena 45(3):227–252. https://doi.org/10.1016/S0341-8162(01)00147-2

Desir G, Marín C (2013) Role of erosion processes on the morphogenesis of a semiarid badland area. Bardenas Reales (NE Spain). Catena 106:83–92. https://doi.org/10.1016/j.catena.2013.02.011

Díaz-Hernández JL, Juliá R (2006) Geochronological position of badlands and geomorphological patterns in the Guadix–Baza Basin (SE Spain). Quat Res 65(3):467–477. https://doi.org/10.1016/j.yqres.2006.01.009

Faulkner H, Alexander R, Teeuw R, Zukowskyj P (2004) Variations in soil dispersivity across a gully head displaying shallow sub-surface pipes, and the role of shallow pipes in rill initiation. Earth Surface Processes Landforms 29(9):1143–1160. https://doi.org/10.1002/esp.1109

Gallart F, Solé A, Puigdefábregas J, Lázaro R (2002) Badland systems in the Mediterranean. In: Bull JL, Kirkby MJ (eds) Dryland rivers: hydrology and geomorphology of semi-arid channels. Wiley, Chichester, pp 299–326

García-Fayos P, Recatala TM (1992) The seed reserves in a badlands basin, Petrer, Alicante. Pirineos 140:29–36. https://doi.org/10.3989/pirineos.1992.v140.170

Harvey A (2004) Badland. In: Goudie AS (ed) Encyclopedia of geomorphology. Routledge, London, pp 45–48

Lázaro R, Calvo-Cases A, Rodríguez-Caballero E, Arnau-Rosalén E, Alexander R, Rubio C, Cantón Y, Solé-Benet A, Puigdefábregas J (2022) Biocrust and catchment asymmetry in Tabernas Desert (Almería, Spain). Geoderma 406:115526. https://doi.org/10.1016/j.geoderma.2021.115526

Martínez-Paz JM, Albaladejo-García JA, Alcon F (2022) When cultural services and biodiversity matter most: gaining a deeper insight into badlands ecosystem services preferences. Land Degrad Dev 34(2):545–557. https://doi.org/10.1002/ldr.4478

Mathys N, Brochot S, Meunier M (1996) Erosion of Terres Noires (Black Earth) in the southern French Alps: a contribution to an assessment of mean annual rates, Draix experimental catchment areas. Revue Geogr Alpine 84(2):17–27

Nadal-Romero E, Martínez-Murillo JF, Vanmaercke M, Poesen J (2011) Scale-dependency of sediment yield from badland areas in Mediterranean environments. Progr Phys Geogr 35(3):297–332. https://doi.org/10.1177/0309133311400330

Nadal-Romero E, Rodríguez-Caballero E, Chamizo S, Juez C, Cantón Y, García-Ruiz JM (2021) Mediterranean badlands: their driving processes and climate change futures. Earth Surface Processes Landforms 47(1):17–31. https://doi.org/10.1002/esp.5088

Peng H (2001) Danxia geomorphology of China: a review. Chin Sci Bull 46(1):38–44. https://doi.org/10.1007/BF03187234

Phillips CP (1998) The badlands of Italy: a vanishing landscape? Appl Geogr 18(3):243–257. https://doi.org/10.1016/S0143-6228(98)00005-8

Stetler LD, Benton R, Weiler MW (2011) Erosion rates from Badlands National Park. In: International symposium on erosion and landscape evolution, Anchorage, AK, p 9

Syvitski JPM, Kettner A (2011) Sediment flux and the Anthropocene. Philos Trans R Soc A 369(1938):957–975. https://doi.org/10.1098/rsta.2010.0329

UNCCD (1994) United Nations convention to combat desertification. United Nations, Paris

Valentin C, Poesen J, Li Y (2005) Gully erosion: impacts, factors and control. Catena 63(2–3):132–153. https://doi.org/10.1016/j.catena.2005.06.001

Zgłobicki W, Poesen J, Cohen M et al (2019) The potential of permanent gullies in Europe as geomorphosites. Geoheritage 11:217–239. https://doi.org/10.1007/s12371-017-0252-1

Chapter 4
The Harms and Benefits of Dust

Sergio Rodríguez, Jessica López-Darias and Paul Ginoux

Abstract Dust particles emitted from arid soils remain airborne for weeks, affecting climate, ecosystems and air quality. Dust may also impact food production and fisheries. In Spain, dust originates mainly from two sources: local agricultural activities and desert-dust transported from North Africa. Agricultural-dust emissions are associated with intensive farming of cereals, almonds, olives and other crops along the basins of the rivers Ebro (in the northeast), Duero (in the northwest) and Guadalquivir (in the south), as well as Murcia, Alicante, Castilla-La Mancha and southern Extremadura. An important hotspot occurs in Córdoba and Jaen, where soil shows severe disturbances by agricultural machinery, breaking the natural geological and biological soil crusts. Emissions of agricultural-dust result in the loss of minerals, nutrients and organic matter, contributing to a soil degradation that be mitigated through vegetation cover. On the other hand, the intensity of North African dust events has experienced a sharp increase. Between 2020 and 2022, Spain experienced record-breaking Saharan dust episodes, with dust concentrations reaching 5254 $\mu g \cdot m^{-3}$ (1-h average) in Gran Canaria (22-Feb-2020) and 3069 $\mu g \cdot m^{-3}$ (24-h average) in Almería (15-Mar-2022). Future dust impacts in Spain will depend on agricultural land management and the extent to which climate change drives Saharan dust transport and the projected northward expansion of North African drylands.

S. Rodríguez (✉) · J. López-Darias
Instituto de Productos Naturales y Agrobiología (IPNA), CSIC, Tenerife, Spain
e-mail: sergio.rodriguez@csic.es

J. López-Darias
e-mail: j.lopez.darias@csic.es

P. Ginoux
NOAA/OAR Geophysical Fluid Dynamics Laboratory, Princeton, NJ, USA
e-mail: Paul.Ginoux@noaa.gov

J. Martínez-Valderrama and J. Olcina Cantos (eds.), *The Labyrinth of Desertification*,
https://doi.org/10.1007/978-3-032-09648-7_4

Keywords Desert dust · Agriculture dust · Aerosols · Aeolian erosion · Dust and fisheries · Air quality

Dust Emissions into the Atmosphere

The action of wind on certain arid terrains results in the emission of dust particles, also known as aeolian dust, mineral dust, or simply dust. The saltation and ballistic impacts that trigger dust emissions may occur at wind speeds as low as 20 km·h^{-1} (Helgren and Prospero 1987) depending on the type of terrain, vegetation cover, and air and soil humidity.

Once in the air, giant dust particles (millimetres) tend to settle near the source region, whereas particles smaller than 100 micrometres (µm) and particularly those under 10 µm can be lifted several kilometres above ground and remain suspended in the atmosphere for about 2 weeks as aerosols (Huneeus et al. 2011). These small aerosol particles may experience long-range transport, potentially travelling hundreds to thousands of kilometres depending on wind speed (Fig. 4.1).

It is estimated that around 4700 million tons of dust are emitted into the atmosphere every year (Kok et al. 2023). The main source regions are located within the so-called 'Dust Belt', which spans North Africa, the Middle East, and interior Asia (Fig. 4.1) (Prospero et al. 2002). Additional aeolian dust sources are found in the subtropical regions of North America, South America, southern Africa, and Australia. Although on a smaller scale, dust emissions also occur at high latitudes, particularly in Arctic deserts such as those in Alaska, Iceland and Patagonia (Fig. 4.2). These dust aerosols influence climate, ecosystems and air quality.

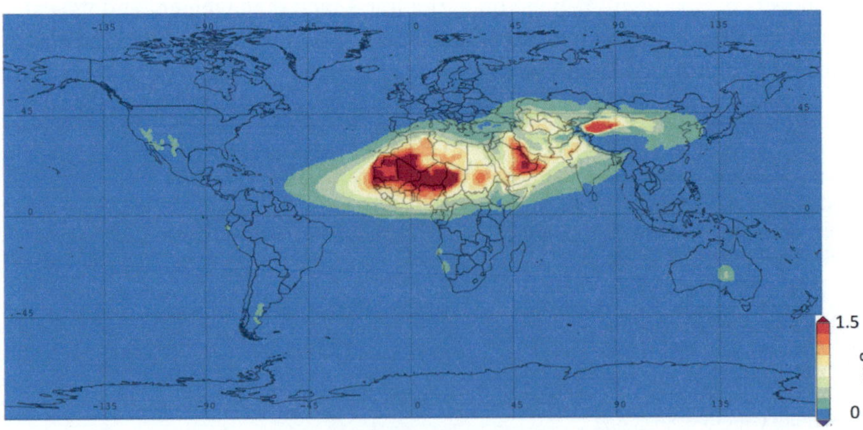

Fig. 4.1 Climatology of the dust load for the period 1990–2020. Illustration based on Giovanni-NASA using the MERRA-2 model (Gelaro et al. 2017)

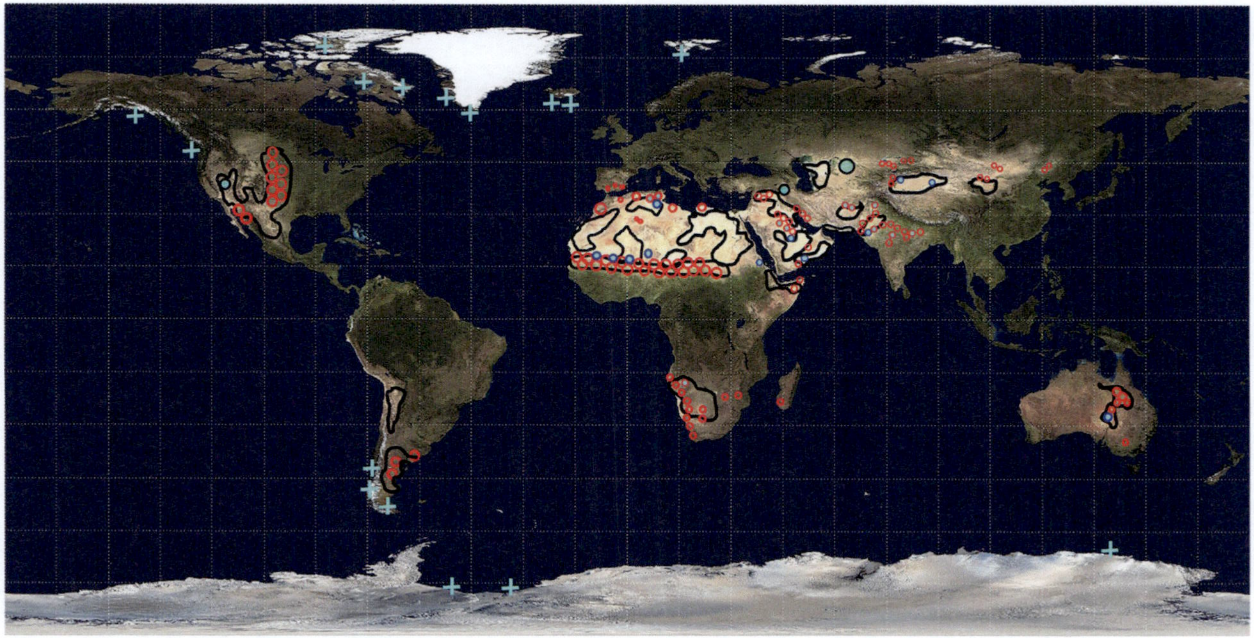

Fig. 4.2 Global distribution of dust source regions (black lines) illustrating sources linked to (i) agricultural land (red circle), (ii) agricultural land near water bodies (red circle filled with light blue), (iii) ephemeral/stationary lakes (blue circle filled with light blue), (iv) large dried-up lakes (Owen-California, Aral-Uzbekistan and Urmia-Iran) and (v) glacial lakes (light blue cross) (Bullard et al. 2016; Ginoux et al. 2012; Prospero et al. 2002)

Dust emissions in Spain are relatively weak compared to other regions; however, Spain receives major dust clouds from Sahara, so the influence of dust emission, desertification and climate change needs to be addressed in a transboundary scale.

Box 4.1: Dust, Climate and Climate Change

Dust and climate influence each other through important feedback mechanisms. Dust aerosols scatter solar and terrestrial radiations and act as ice nuclei catalysts on the formation of ice clouds as cirrus clouds. Dust deposition in the ocean provides iron, phosphorus and other nutrients to phytoplankton, increasing primary productivity and acting as a carbon dioxide sink. The global net effect of dust aerosols is to cool the atmosphere, although it does not offset the warming caused by excess greenhouse gases. Radiative forcing due to direct dust-radiation interactions is estimated at $-0.2\,W\ m^{-2}$. Studies based on marine sediments indicate that dust inputs to the ocean could explain between 40 and 50 ppm of the observed CO_2 variations between glacial and interglacial periods over the last million years (Martínez-Garcia et al. 2009).

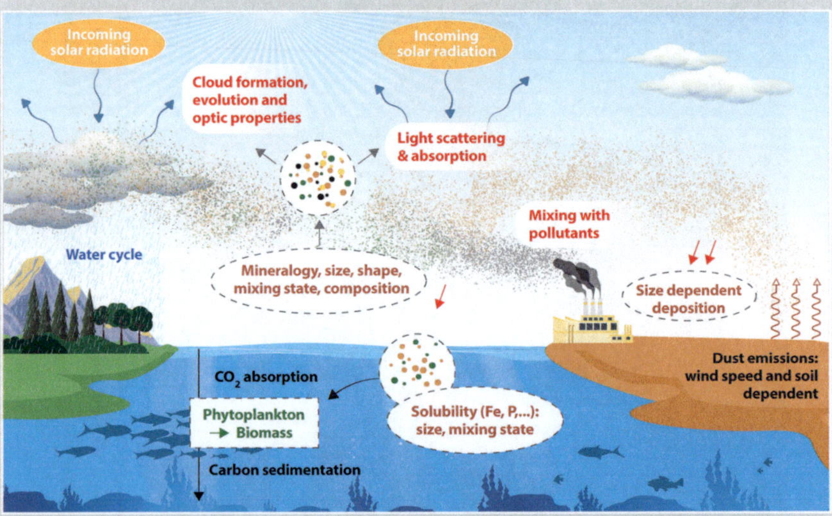

Illustration on how dust influences the climate (Source: adapted from Rodríguez et al. 2012)

The dust load in the atmosphere has experienced a significant increase since the development of industrial agriculture, reaching a peak in the 1980s (80% higher than the pre-industrial period), and a subsequent reduction in emissions is attributed to the slowdown in atmospheric circulation linked to climate change (Evan et al. 2016). Currently, the dust load is 56% higher than in the pre-industrial period (Kok et al. 2023). Recent climate projections point to an expansion of North African arid zones towards mainland Spain and a shift in dust transport routes northward (Liu et al. 2024).

Origin of Aeolian Dust

Natural Dust

Subtropical regions include dust sources that are considered natural, as it is considered that they have not been disturbed by human activities. The most active dust sources are located in topographic lows accumulating fluvial and lacustrine sediments of ancient rivers and lakes (Prospero et al. 2002). Some of them are located in North Africa, in current hyper arid regions (aridity index <0.05; annual precipitation <200 mm) that held large lakes and rivers in the last 'Green Sahara' period (5000–15,000 years ago) linked to the more abundant rainfalls linked to the monsoon. This type of large dust-sources occurs in several subtropical regions, as Rub'al Khali (Saudi Arabia) or the Taklamakan desert (China). Saharan dust aerosols transported nowadays over the Atlantic are mixed with skeletons of (few microns size) fresh water diatoms, reminiscent of a wetter past (Barkley et al. 2021). Currently, ephemeral lakes, such as those characteristics of arid regions with monsoon climates, continue accumulating sediments acting as dust sources with significant contribution to dust load especially in the southern hemisphere (Song and Ginoux 2025).

Understanding the processes that led to the formation of these natural sources is crucial to preventing the creation of new anthropogenic sources of dust.

Anthropogenic Dust

About 25 million tons of dust blows in the atmosphere every day, so-called dust load (Fig. 4.1). Studies based on modelling constrained with paleorecords indicate that the current dust load (2020s) is 56% higher than in the pre-industrial era (<1850) (Kok et al. 2023). Paleo records indicate that this increase in dust load is linked to (i) the expansion of industrial agriculture since the development of fertilizers in the nineteenth century (red circles in Fig. 4.2) and (ii) the desiccation of lakes due to agricultural irrigation, as occurred in the Aral Sea Lake (Uzbekistan) or Lake Urmia (Iran), or for urban supply, as occurred with Lake Owen (USA) at the beginning of the twentieth century (light blue filled circles in Fig. 4.2). The increase in the atmospheric dust load has been of about 47% in North Africa, 76% in Asia and 27% in the Southern Hemisphere (Kok et al. 2023). Currently, North Africa emits ~50% of the dust emitted globally, a fifth of this value (~10% of the global) is emitted in areas of the Sahel currently dedicated to agriculture (red circles, Fig. 4.2), and where paleo records indicate that dust emissions increased dramatically with the expansion of commercial agriculture in the nineteenth century. Studies based on satellite observations estimate that currently, ~25% of the dust emitted globally is of anthropogenic origin, linked to agricultural soil (red dots, Fig. 4.2), and secondarily, to the drying up of lakes (blue filled circles, Fig. 4.2) (Ginoux et al. 2012).

Box 4.2: Dust, Ecosystems and Fisheries

Emissions, transport and deposition of dust have a huge impact on ecosystems, with implication on food production. Dust emissions lead to loss of minerals, nutrients and organic matter from the soil, contributing to soil depletion. Dust deposition on leaves also reduces plants' photosynthetic activity. Agriculture is the main source of anthropogenic dust; dust storms, in turn, have a huge impact on food production (Middleton 2024). Conversely, atmospheric deposition of dust in the open ocean provides nutrients to the marine ecosystem, influencing the biogeochemical cycles of iron, phosphorus, manganese, copper, zinc and nickel, among other nutrients.

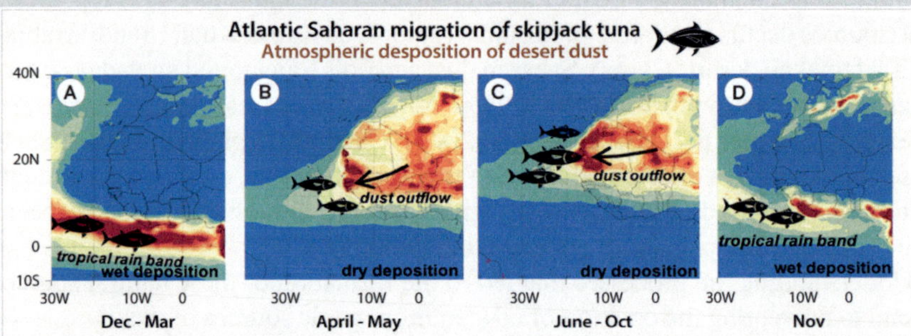

Seasonal migration of skipjack tuna, illustrating how the main fishing areas of skipjack tuna are located in waters affected by the massive deposition of Saharan dust (Rodríguez et al. 2023)

The atmospheric deposition of North African dust in the Atlantic supports the growth of phytoplankton, creating new organic matter that is distributed across the marine ecosystem's food web until it reaches top predators. This process has implications for the fishing industry. The main fishing areas of skipjack tuna (*Katsuwonus pelamis*) occur in water affected by massive dust deposition (Rodríguez et al. 2023). From January to August, skipjack tuna performs a seasonal migration from the equatorial Atlantic to the waters off Mauritania to the Canary Islands, tracking the seasonal shift of Saharan dust deposition (Fig. 4.1).

Agricultural Dust Emissions in Spain

Satellite observations evidence that mainland Spain is affected by agriculture dust emissions in several regions, mainly in Castilla y León, Aragón, Navarra, Castilla-La Mancha, Extremadura, Andalucía, Murcia and the southern part of the Comunidad Valenciana (Fig. 4.3). Emissions are important in the intensive farming along the basins of the rivers Ebro, Duero and Guadalquivir. In these

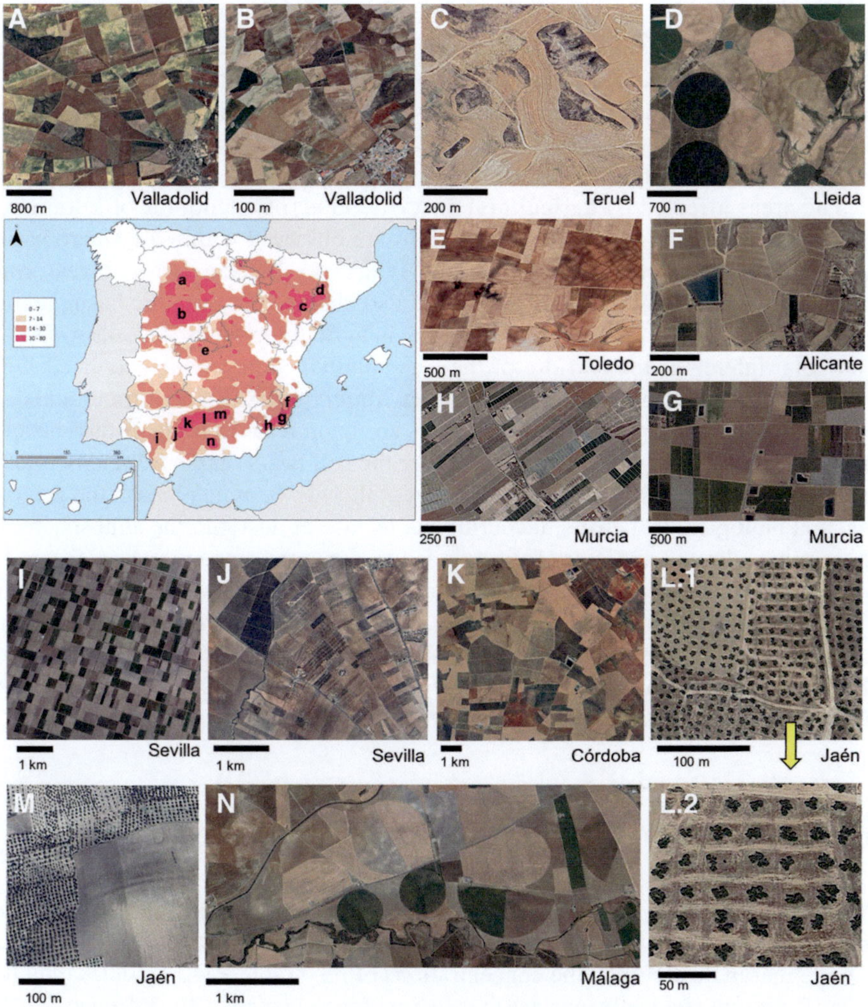

Fig. 4.3 Number of days per year which the agricultural dust load exceeds the threshold of 0.2 aerosol optical depth according to satellite observations (Ginoux et al. 2012). Examples (Google Earth™) of dust-emitting areas are shown, indicating nearby locations: **A** Valladolid (Cuenca de Campos), **B** Valladolid (Lomoviejo), **C** Teruel (Híjar), **D** Lleida (Gimenells), **E** Toledo (Villatobas), **F** Alicante (Alicante), **G** Murcia (La Palma), **H** Murcia (Lorca), **I** Seville (Lebrija), **J** Seville (Osuna), **K** Córdoba (La Carlota), **L** Jaén (Lopera), **M** Jaén (Calzadilla) and **N** Málaga (Antequera)

regions, the dust load exceeds the threshold of 0.2 aerosol optical depth (a measure of the attenuation of sunlight by suspended particles) between 7 and 80 days per year (Fig. 4.3). These exceedances are much lower than those typically detected in North Africa (Ginoux et al. 2012). A satellite view evidences that these dust emissions are mostly linked to agricultural lands of olive, almond, cereal and vegetables, among other crops.

The areas affected by agricultural dust extend across hundreds of kilometres (Fig. 4.3). In Castilla y León, dust emissions are observed within the Duero basin, across the provinces of Valladolid, Segovia, Avila and Salamanca, where cereal crops cultivation is abundant. Dust emissions are also observed in the Ebro basin, across Navarra, Aragon and Eastern Catalonia, also in agriculture fields (Fig. 4.3A–D). Of particular interest is the corridor of the Guadalquivir Valley, which connects with regions of Eastern Spain as Albacete, Cuenca, Murcia and Alicante. Agricultural dust emissions appear in lands without vegetation, i.e., unprotected terrains. A dust-hotspot is observed across Cordoba and Jaen, where the soil bears marks from agricultural machinery (e.g., tractors, plough, etc.) (Fig. 4.3L.1, L.2), which breaks the geological and biological soil crusts, favouring the release of dust into the atmosphere. In these olive, almond and other fruit crop lands, dust emissions may be reduced by vegetation land coverage, e.g., by using practices such as those recommended in the 'Olivares Vivos' project (Tarifa et al. 2021). It is noteworthy that large hotspots are not observed in mining areas, as this industry is subject to strict environmental regulations that minimize dust emissions. Nor are they observed in Almería, where much of the agricultural activity is carried out in greenhouses.

Extreme Saharan Dust Events Hits Spain

North African dust is regularly transported to the Atlantic in the Saharan Air Layer airstream (Prospero et al. 2021), which often deviates towards Europe (Fig. 4.1). During these Saharan dust events, the concentrations of respirable PM_{10} particles (smaller than 10 microns) typically increase to values within the range of 40–300 $\mu g\ m^{-3}$ in the Canary Islands and 40–100 $\mu g\ m^{-3}$ in mainland Spain (24-h averages). This dynamic of 'regular dust events' changed in the period 2020 and 2022, when extreme Saharan dust events hit Spain (Fig. 4.4) (Rodríguez and López-Darias 2024). During these extreme dust events, record breaking PM_{10} concentrations were measured in the air quality networks, with hourly and daily averages concentrations reaching 5254 and 4605 $\mu g\ m^{-3}$ (1-h average) and maximums of 1840 and 3069 $\mu g\ m^{-3}$ (24-h average) in Gran Canaria (February 22, 2020) and Almería (March 15, 2022) (Rodríguez and López-Darias 2024), respectively.

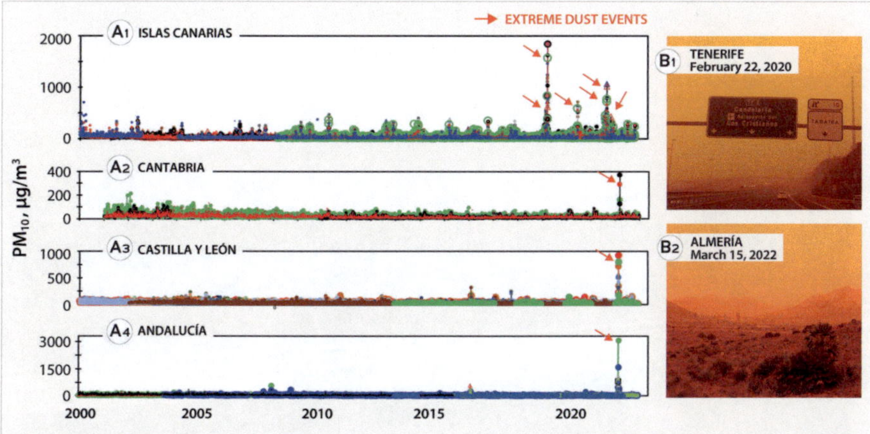

Fig. 4.4 Extreme Saharan dust events. Historical record of respirable PM_{10} particles (smaller than 10 microns) measured in the air quality networks of the Canary Islands, Cantabria, Castilla y León and Andalusia, indicating with red arrows the 6 extreme dust events recorded in the period 2020–2022. Photographs taken in Tenerife (B1) and Almería (2) during the extreme dust events that affected these territories in the periods 22–29 February 2020 and 15–16 March 2022 (Rodríguez and López-Darias 2024)

Box 4.3. Dust and Air Quality

The inhalation of dust blowing in the ambient air during African dust events has adverse effects on the cardiorespiratory system. Increases of 10 $\mu g \cdot m^{-3}$ in the dust concentrations in ambient air are associated with (Dominguez-Rodriguez et al. 2020, 2021):

- Inflammation of the respiratory air ways, as indicated by the increase in the concentrations of biomarkers of inflammatory processes in the sputum of patients, such as hydroxyproline and transforming growth factor (TGF-b1).
- A 2% increase in the risk of cardiovascular death (per every 10 $\mu g \cdot m^{-3}$ of dust). In the case of extreme dust events recorded in Spain, when PM_{10} levels exceeded 3000 $\mu g \cdot m^{-3}$ (24-h average), the cardiovascular mortality risk would be greater than 600%. During a period of 4 years, the 86% of in-hospital heart-failure mortality cases recorded in the Canary Islands University Hospital occurred in patients admitted during Saharan dust events.

People using masks during an extreme Saharan dust in Lanzarote. Photo by Sergio Rodríguez

References

Barkley AE, Olson NE, Prospero JM et al (2021) Atmospheric transport of North African dust-bearing supermicron freshwater diatoms to South America: implications for iron transport to the equatorial North Atlantic Ocean. Geophys Res Lett 48:e2020GL090476. https://doi.org/10.1029/2020GL090476

Dominguez-Rodriguez A, Rodríguez S, Baez-Ferrer N et al (2020) Impact of Saharan dust exposure on airway inflammation in patients with ischemic heart disease. Transl Res 224:16–25. https://doi.org/10.1016/j.trsl.2020.05.011

Domínguez-Rodríguez A, Báez-Ferrer N, Abreu-González P et al (2021) Impact of desert dust events on the cardiovascular disease: a systematic review and meta-analysis. J Clin Med 10(4):727. https://doi.org/10.3390/jcm10040727

Evan AT, Flamant C, Gaetani M, Guichard F (2016) The past, present and future of African dust. Nature 531:493–495. https://doi.org/10.1038/nature17149

Bullard JE, Baddock M, Bradwell T et al (2016) High-latitude dust in the Earth system. Rev Geophys 54:447–485. https://doi.org/10.1002/2016RG000518

Gelaro R, McCarty W, Suárez MJ et al (2017) The modern-era retrospective analysis for research and applications, version 2 (MERRA-2. J Clim 30:5419–5454. https://doi.org/10.1175/JCLI-D-16-0758.1

Ginoux P, Prospero JM, Gill TE et al (2012) Global-scale attribution of anthropogenic and natural dust sources and their emission rates based on MODIS Deep Blue aerosol products. Rev Geophys 50: https://doi.org/10.1029/2012RG000388

Helgren DM, Prospero JM (1987) Wind velocities associated with dust deflation events in the Western Sahara. J Clim Appl Meteorol 26:1147–1151. https://doi.org/10.1175/1520-0450(1987)026<1147:Wvawdd>2.0.Co;2

Huneeus N, Schulz M, Balkanski Y, Griesfeller J, Prospero J, Kinne S, Bauer S, Boucher O, Chin M, Dentener F, Diehl T, Easter R, Fillmore D, Ghan S, Ginoux P, Grini A, Horowitz L, Koch D,

Krol MC, Landing W, Liu X, Mahowald N, Miller R, Morcrette J-J, Myhre G, Penner J, Perlwitz J, Stier P, Takemura T, Zender CS (2011) Global dust model intercomparison in AeroCom phase I. Atmos Chem Phys 11:7781–7816. https://doi.org/10.5194/acp-11-7781-2011
Kok JF, Storelvmo T, Karydis VA et al (2023) Mineral dust aerosol impacts on global climate and climate change. Nat Rev Earth Env 4:71–86. https://doi.org/10.1038/s43017-022-00379-5
Liu J, Wang X, Wu D, Wei H, Li Y, Ji M (2024) Historical footprints and future projections of global dust burden from bias-corrected CMIP6 models. NPJ Clim Atmos Sci 7(1):1. https://doi.org/10.1038/s41612-023-00550-9
Martínez-Garcia A, Rosell-Melé A, Geibert W et al (2009) Links between iron supply, marine productivity, sea surface temperature, and CO2 over the last 1.1 Ma. Paleoceanography 24: https://doi.org/10.1029/2008PA001657
Middleton N (2024) Impacts of sand and dust storms on food production. Environ Res Food Syst 1(2):022003. https://doi.org/10.1088/2976-601X/ad63ac
Prospero JM, Ginoux P, Torres O et al (2002) Environmental characterization of global sources of atmospheric soil dust identified with the Nimbus 7 Total Ozone Mapping Spectrometer (TOMS) absorbing aerosol product. Rev Geophys 40(1):2–31. https://doi.org/10.1029/2000RG0000952
Prospero JM, Delany AC, Delany AC, Carlson TN (2021) The discovery of African dust transport to the Western Hemisphere and the Saharan air layer: a history. Bull Am Meteorol Soc 102:E1239–E1260. https://doi.org/10.1175/BAMS-D-19-0309.1
Rodríguez S, Alastuey A, Querol X (2012) A review of methods for long term in situ characterization of aerosol dust. Aeolian Res 6:55–74. https://doi.org/10.1016/j.aeolia.2012.07.004
Rodríguez S, Calzolai G, Chiari M et al (2020) Rapid changes of dust geochemistry in the Saharan Air Layer linked to sources and meteorology. Atmos Env 223:117186. https://doi.org/10.1016/j.atmosenv.2019.117186
Rodríguez S, Riera R, Fonteneau A, Alonso-Pérez S, López-Darias J (2023) African desert dust influences migrations and fisheries of the Atlantic skipjack-tuna. Atmos Env 312:120022. https://doi.org/10.1016/j.atmosenv.2023.120022
Rodríguez S, López-Darias J (2024) Extreme Saharan dust events expand northward over the Atlantic and Europe, prompting record-breaking PM 10 and PM 2.5 episodes. Atmos Chem Phys 24:12031–12053. https://doi.org/10.5194/acp-24-12031-2024
Song Q, Ginoux P (2025) The contribution of ephemeral lakes to global dust cycle and direct radiative effect. Geophys Res Lett 52:e2025GL116689. https://doi.org/10.1029/2025GL116689
Scheuvens D, Schütz L, Kandler K, Ebert M, Weinbruch S (2013) Bulk composition of northern African dust and its source sediments — a compilation. Earth Sci. Rev. 116:170–194. https://doi.org/10.1016/j.earscirev.2012.08.005
Tarifa R, Martínez-Núñez C, Valera F et al (2021) Agricultural intensification erodes taxonomic and functional diversity in Mediterranean olive groves by filtering out rare species. J Appl Ecol 58(10):2266–2276. https://doi.org/10.1111/1365-2664.13970

Chapter 5
Shrub Encroachment in Drylands: Scientific Basis Regarding Its Positive and Detrimental Effects

Santiago Soliveres

Abstract Shrub encroachment, the increase in abundance or cover of woody species in areas where they were formerly absent or really scarce is a very contested process regarding its consequences for ecosystem functioning. Some scientists argue that shrub encroachment increase heterogeneity and concentrate scarce resources (water and nutrients) beneath them, therefore compromising plant growth (and with it forage availability and soil protection). Others, instead, see this process as a beneficial recovery from disturbances, with clear benefits on biodiversity, carbon sequestration and overall ecosystem health. In this chapter, I revisit the extend of shrub encroachment as a global phenomenon, which shows some particular characteristics in Spain. I also show in which context (in terms of climate, land use and plant cover) and for which ecological properties are we to expect either positive responses or detrimental effects from the encroachment of woody plants. I pay particular attention to the linkages between shrub encroachment and three widely use indicators of desertification, namely soil organic carbon, soil erosion and plant primary productivity.

Keywords Dryland · Biodiversity · Plant productivity · Soil carbon · Water · Infiltration · Woody · Encroachment

S. Soliveres (✉)
Department of Ecology, University of Alicante, Alicante, Spain
e-mail: santiago.soliveres@ua.es

Multidisciplinary Institute for Environmental Studies "Ramon Margalef", University of Alicante, Alicante, Spain

J. Martínez-Valderrama and J. Olcina Cantos (eds.), *The Labyrinth of Desertification*, https://doi.org/10.1007/978-3-032-09648-7_5

Scrublands and Brownfields: An Ambiguous Landscape of Desertification

The debate between the benefits of shrub encroached lands or the resulting costs (economic and ecological) is nothing new. The "Degraded Scrubland and Brownfields" landscape harbors this paradox, and it is therefore appropriate to highlight the elements of this debate in the Spanish context. The environmental costs range from the deterioration of grazing areas (where shrublands are often classified as "green deserts") to soil deterioration as a result of the spread of shrubs into former grasslands. Here, we can see how the terminology biases the discourse, because speaking of spread, which resembles disease, is not the same as speaking of restoration, where shrubs stand as a hallmark of the recovery of vegetation cover, using species native to drylands instead of larger, faster-growing tree species. The Spanish National Programme Against Desertification (PAND, the Spanish acronym) refers to "the large amount of land that supports wasteland or shrubland formations resulting from the degradation of more mature vegetation formations, endowed with greater ecological complexity." Ultimately, discussing shrub encroachment and desertification requires several clarifications, which I hope to properly address in this chapter.

Shrub Encroachment, a Global Phenomenon... Also in Spain?

Shrub encroachment is defined as the increase in shrubland cover in areas commonly occupied by pastures, crops, or steppes. Shrub encroachment has been classified as a global phenomenon, with increases in the area occupied by shrubs of between 0.5 and 2% additional surface area/year, recorded in places as diverse as Argentina, Australia, the United States, and South Africa, to name a few (Eldridge et al. 2011; Anadón et al. 2014). Does this phenomenon also occur in Spain? Well, it depends. If we look at forest area (woody species in general, including trees and shrubs), the answer is a resounding yes; this is mainly due to the abandonment of extensive grazing, crops, and fuel extraction, as well as numerous reforestation projects carried out since the 1960s. If we refer only to non-wooded forest area (i.e., shrub-dominated areas), this is a different story. According to the Spanish Society of Forest Sciences (Bravo et al. 2017), forest area in our country increased by 25% between 1990 and 2015, but not the treeless woody area, which decreased by 30% over the same period. The fourth Spanish Forest Inventory provides similar data: forest area has increased by approximately 20% (11.8 vs. 14.7 million Ha between 1965–1975 and 2008–2017), while the treeless woody area has decreased by a similar percentage (13.8 million Ha vs. the current 11.6 million Ha).

While there does seem to be an overall expansion of the area occupied by shrubs in our country, the fact is that the area occupied by these short woody species is not small (somewhere between 13 and 24% of our territory, according to different sources; Bravo et al. 2017; Instituto Geográfico Nacional 2018). Indeed, in some regions of

Fig. 5.1 Shrubs colonizing stressful environments, such as salinized, poor or heavily eroded soils. Tabernas desert, Almeria, Spain. Photo by Jaime Martínez-Valderrama

Spain shrub-dominated areas have increased in recent decades (up to 37% more treeless wooded area in Catalonia between 2007 and 2017). Therefore, it is common in our country for lands formerly occupied by crops, steppes, or grasslands to see an increase in shrub cover. It is also not uncommon for forest areas that have been burned frequently (enough to eliminate the tree layer) to now be occupied by shrubs (Baeza et al. 2007). In general, the increase in the abundance of shrubland in our territory seems to be favored on steep slopes, those more exposed to solar radiation, or on less fertile soils (Fig. 5.1).

Clarifying Concepts: What Does Shrub Encroachment Entail in Different Ecological Contexts?

While the process is clear, agreement regarding the environmental consequences of shrub encroachment is less well-established. This is perhaps because such consequences depend on which shrub species are colonizing, where they are doing so (climatic conditions, previous uses), or the ecosystem attribute being compared. An increase in the cover of a shrub species that fixes atmospheric nitrogen or whose canopy protects the soil is not the same as an increase in cover of a shrub species that emits allelopathic compounds and whose leaves are separated from the soil surface by V-shaped stems (Eldridge et al. 2011). Nor is an increase in shrub cover in a dry area, or with discontinuous cover (steppes, cultivated land), the same as an increase in grasslands with cover close to 100% or in more humid areas (Maestre et al. 2009; Soliveres et al. 2014). And it is not the same if what interests us is the forage value of the area, than if we are interested in the sequestration of atmospheric CO_2 by vegetation, or the richness of bird or plant species (Eldridge et al. 2011; Anadón et al. 2014). Perhaps the clearest consequences of shrub encroachment, and those least dependent on the ecological context, are: (1) a decrease in forage value, especially in areas with active grazing, and (2) an increase in atmospheric CO_2 sequestration, although not necessarily an increase in primary productivity. If we focus on forage value for livestock production, available estimates indicate that for every additional 1% in shrub cover, forage productivity declines by 2.5% (Anadón et al. 2014), partly explained by a reduction in the abundance of grasses and legumes, the basis of extensive livestock feed (Gómez-García et al. 2023).

If we are more interested in atmospheric CO_2 capture, this will increase as the cover of shrubs or other woody species increases. Available estimates range from 0.28 to 2.97 new tons of plant biomass (dry matter ha^{-1} $year^{-1}$), depending on the type of shrub formation (0.83 t ha^{-1} $year^{-1}$ in thyme scrublands, known as "degraded scrublands and wastelands" according to forestry terminology). These rates of accumulation of new plant biomass are accompanied by an increase of between 3.2 and 4 new tons of organic carbon incorporated annually into the soil per hectare (Montero et al. 2020). Therefore, shrub encroachment generally has positive effects on the amount of carbon stored, both aboveground and in the soil (Maestre et al. 2009; Eldridge et al. 2011; Zhou et al. 2017). It's worth remembering here, on the one hand, that (low) soil carbon content is one of the variables frequently used as indicator of desertification. On the other hand, much of the biomass accumulated by these shrublands has a low moisture content, which can facilitate the spread of wildfires (see Chap. 6) (Fig. 5.2).

Another commonly used indicator of desertification (changes in primary productivity) does not appear to follow any clear trend with shrubland expansion, with 19 studies showing positive effects of this process, 20 showing negative effects, and 9 studies showing no change (Eldridge et al. 2011). These less evident effects on primary productivity also extend to those recorded on species richness of different groups of organisms (birds, vascular plants, or insects), or soil erosion rates, as these are generally more dependent on the ecological context. The effects on diversity tend to be more positive in more arid environments (Eldridge et al. 2011; Anadón et al. 2014), under intermediate dominance of shrubs in the community (occupying 40–60% of the total vegetation cover; Sirami et al. 2009; Soliveres et al. 2014), in "dehesas" (savannahs) or ecosystems with discontinuous cover than in grasslands with continuous cover (Maestre et al. 2009; López-Díaz et al. 2015; Stanton et al. 2021) and for shrub species that are short, with the canopy associated with the soil and/or nitrogen fixers (Eldridge et al. 2011). Although much less studied, particularly in our country, soil erosion and loss processes (also associated with desertification) appear to increase when ecosystems previously occupied by grasslands with high cover are encroached by shrubs, increasing the proportion of bare soil (Manjoro et al. 2012). These effects are obviously less negative in areas with scarce soil cover (deforested areas, former

Fig. 5.2 Shrublands protect soils from erosion, favor water infiltration and retention (and increase underground water storage), and enhance organic matter inputs into the soil. Yet, these are often considered barren or unproductive territories as opposed to the intensively managed olive orchards sometimes replacing them (in the back). The latter are often considered productive and desirable despite they may come at a cost of depauperated underground water reservoirs (if irrigated), lower biodiversity levels and eroded soils (ecological attributes that, despite their clear importance, are often not included in economic balances), if not managed sustainably. Photo by Jaime Martínez-Valderrama

croplands), which commonly have high rates of soil erosion (see Chap. 15), which decrease drastically once colonized by shrubs (Lasanta et al. 2019).

Finally, an important aspect to consider in the relationship between shrub encroachment and desertification concerns water balances. This topic has been relatively little studied (Acharya et al. 2018), although the depletion of groundwater reserves is far from being a minor issue in the field of desertification in our country. In general, higher organic matter contents under the canopy of shrublands increase water infiltration and retention into the soil after rainfall events (e.g., Berdugo et al. 2014). However, the canopies of these same shrubs also intercept large amounts of water, particularly during small rainfall events (<20 mm), preventing water from reaching the soil and thus hindering the recharge of groundwater reserves after these small rainfall events (Moghli, Master's thesis, unpublished). Overall, existing evidence suggests that spaces covered with shrublands generate greater water flows into underground reservoirs than those occupied by trees. However, comparisons between shrublands vs. grasslands, both in terms of groundwater recharge and use, suggest that shrubland ecosystems generate lower recharge flows than those without the presence of woody species (Acharya et al. 2018).

Box 5.1: Shrub Encroachment and Desertification? Unclear Linkages

While tree cover is increasing in our country, shrubland coverage is not. However, shrublands occupy a significant proportion of our territory (13–24%), and are increasing in some regions. Shrub encroachment generally has a positive effect in areas with low and heterogeneous vegetation cover, with little initial shrubland dominance, and low rainfall, characteristics commonly found in arid and semi-arid areas. These benefits of shrubland expansion are particularly evident on biodiversity, the ability to sequester atmospheric CO_2, soil fertility, and water infiltration capacity.

Esparto grasslands in the southeastern Iberian mountain ranges, a type of vegetation well adapted to the aridity of this region. Photo by Jaime Martínez-Valderrama

This is why the introduction of shrublands (especially resprouting shrubs with a low canopy associated to the soil and with zoochorous dispersal) has been used to restore degraded areas in semi-arid regions. Conversely, in areas with continuous vegetation cover, higher rainfall, or of pastoral value, shrub encroachment can lead to declines in forage productivity, a greater risk of fire, or reduced aquifer recharge, depending on the type of cover being compared. Following criteria based on soil erosion rates, changes in primary productivity or soil organic carbon content, all important indicators of desertification processes, we found scarce evidence that shrub encroachment increase desertification in our country. The latter is particularly true in arid and semi-arid environments, with discontinuous vegetation cover, which are precisely the most prone to desertification.

References

Acharya BS et al (2018) Woody plant encroachment impacts on groundwater recharge. A Rev Water 10(10):1466. https://doi.org/10.3390/w10101466

Anadón JD, et al (2014) Effect of woody-plant encroachment on livestock production in North and South America. Proc Natl Acad Sci USA 111(35):12948–12953. https://doi.org/10.1073/pnas.1320585111

Baeza MJ et al (2007) Human disturbance and environmental factors as drivers of long-term post-fire regeneration patterns in Mediterranean forests. J Vegetation Sci 18:243–252. https://doi.org/10.1111/j.1654-1103.2007.tb02535.x

Berdugo M et al (2014) Vascular plants and biocrusts modulate how abiotic factors affect wetting and drying events in drylands. Ecosystems 17:1242–1256. https://doi.org/10.1007/s10021-014-9790-4

Bravo F et al (2017) La situación de los bosques y el sector forestal en España – ISFE 2017

Eldridge DJ et al (2011) Impacts of shrub encroachment on ecosystem structure and functioning: towards a global synthesis. Ecol Lett 14:709–722. https://doi.org/10.1111/j.1461-0248.2011.01630.x

Gómez-García D et al (2023) Shrub encroachment in Mediterranean mountain grasslands: rate and consequences on plant diversity and forage availability. J Vegetation Sci 34:e13174

Instituto Geográfico Nacional (2018) Biogeografía. In: Atlas Nacional de España. Centro Nacional de Información Cartográfica, Madrid, pp 116–133

Lasanta T et al (2019) Soil degradation, restoration and management in abandoned and afforested lands. Adv Chem Pollut Environ Manag Prot 4:71–117

López-Díaz ML et al (2015) Shrub encroachment of Iberian dehesas: implications on total forage productivity. Agroforest Syst 89:587–598. https://doi.org/10.1007/s10457-015-9793-y

Maestre FT et al (2009) Shrub encroachment can reverse desertification in semi-arid Mediterranean grasslands. Ecol Lett 12:930–941. https://doi.org/10.1111/j.1461-0248.2009.01352.x

Manjoro M (2012) Trends in soil erosion and woody shrub encroachment in Ngqushwa District, Eastern Cape Province, South Africa. Environ Manag 49:570–579. https://doi.org/10.1007/s00267-012-9810-0

Montero G et al (2020) Producción de biomasa y fijación de carbon por los matorrales españoles y por el horizonte orgánico superficial de los suelos forestales. Ministerio de Agricultura, Pesca y Alimentación, Madrid

Sirami C et al (2009) The impact of shrub encroachment on savanna bird diversity from local to regional scale. Divers Distrib 15:948–957. https://doi.org/10.1111/j.1472-4642.2009.00612.x

Soliveres S et al (2014) Woody dominance and ecosystem functioning. Glob Ecol Biogeogr 23:1408–1416. https://doi.org/10.1111/geb.12215
Stanton RA et al (2021) The effects of shrub encroachment on bird occupancy vary with land use in an African savannah. Animal Cons 24:194–205. https://doi.org/10.1111/acv.12620
Zhou Y et al (2017) Soil carbon response to woody plant encroachment: importance of spatial heterogeneity and deep soil storage. J Ecol 105:1738–1749. https://doi.org/10.1111/1365-2745.12770

Chapter 6
Wildfires and Desertification: The Role of Fire Regime

Manuel Esteban Lucas-Borja

Abstract Wildfires are intrinsic ecological processes in many ecosystems, especially in Mediterranean landscapes, where vegetation and soils have evolved under specific fire regimes. However, anthropogenic influences, such as rural depopulation, the decline of agricultural activity, climate change, and fire suppression policies, have transformed natural fire regimes into unnatural ones. These altered regimes often involve increased fire severity and frequency, leading to cascading environmental effects. This article analyzes the link between fire regimes and desertification, emphasizing that it is not fire per se but the nature of the fire regime that determines its ecological impact. Unnatural regimes contribute to soil degradation, loss of vegetation structure, and reduced ecosystem resilience, creating feedback loops that exacerbate desertification. The paper discusses the direct and indirect effects of wildfires on soil, water, air, and vegetation, illustrating how recurrent and intense fires in sensitive ecosystems lead to long-term degradation. Ultimately, effective landscape and fire management strategies must recognize and restore natural fire regimes to prevent irreversible changes and maintain ecosystem services.

Keywords Wildfire regime · Climate change · Mediterranean ecosystems · Soil degradation · Anthropogenic impact

M. E. Lucas-Borja (✉)
Escuela Técnica Superior de Ingeniería Agronómica, de Montes y Biotecnología, Universidad de Castilla La Mancha, Albacete, Spain
e-mail: manuelesteban.lucas@uclm.es

J. Martínez-Valderrama and J. Olcina Cantos (eds.), *The Labyrinth of Desertification*,
https://doi.org/10.1007/978-3-032-09648-7_6

Forest Fires

Wildfires are a key element in many terrestrial ecosystems, as they are considered a fundamental part of different processes and successional dynamics in the landscape (Pausas 2024). Since the Carboniferous period, 307–359 million years ago, this disturbance has acted as a natural phenomenon shaping landscapes and the behaviur of different plant species. Over time and with exposure to fire, plants have adapted to coexist with fire (Resco de Dios 2020) (Photos 1 and 2). This coexistence has developed under specific characteristics defining both fires and particular ecosystems and environments in specific locations, known as the fire regime (Pausas 2024) (Figs. 6.1 and 6.2).

Among the main characteristics that determine a fire regime are seasonality (the time of year the fire occurs), intensity (the amount of energy released by the fire), fire severity (the magnitude of ecological change caused by fire in the ecosystem), recurrence (the time elapsed between wildfires affecting a given area and ecosystem), and the type and extent of the fire (Sayedi et al. 2024). While in the past, different ecosystems and species evolved under what would be considered a natural fire regime, today, due to various circumstances (rural depopulation and the disappearance of the primary sector in these areas, lack of forest management policies and promotion of fire suppression policies, and climate change, among others), the natural fire regime has been altered, turning it into an unnatural or anthropogenic fire regime (Turco et al. 2018).

Fig. 6.1 Regeneration of Aleppo pine one year after the fire in Sierra de los Donceles in 2012 (Hellín, Albacete, Castilla-La Mancha, Spain). Photo by Manuel E. Lucas-Borja

Fig. 6.2 Sprouting of mastic shrub two months after the fire in Agramon in 2020 (Albacete, Castilla-La Mancha, Spain). Photo by Manuel E. Lucas-Borja

The combination of climate change, droughts, biomass accumulation, and an increase in ignition sources has created the perfect conditions for fires to be considered agents contributing to the loss of soil hydrological function and biological productivity (Neary 2018). This fact is fundamental to understanding the relationship between desertification and wildfires, which was one of the situations identified in the 2008 National Action Program as a case of desertification, framed in the "landscape" dedicated to the forestry sector, the "degraded wastelands and shrublands." Recently, the 2022 National Strategy for Combating Desertification has reassessed these landscapes, linking wildfires to the abandonment of agricultural land or the lack of forest ecosystem management. In both cases, the accumulation of flammable material creates landscapes that are more prone to fire, which can subsequently lead to erosion episodes or irreversible changes in the plant community.

Many distinctive characteristics of our landscapes result from human activity (Shakesby 2011). Currently, throughout our geography, there is a noticeable increase in the vertical and horizontal continuity of forest biomass, with a consequent decrease in discontinuities at the landscape level and the absence of mosaics with different land uses that would serve as natural barriers against fire spread (Moreno et al. 2023; Pausas and Fernández-Muñoz 2011). In this context, considering wildfires as the sole or most important agents triggering soil degradation and fertility loss, and thus generating desertification processes, is risky, especially on a broad temporal scale. For example, Pausas et al. (2008) demonstrated that the deterioration of environmental conditions and erosion after wildfires is not a general rule but can be significant when combined with previous human disturbances. Along these lines, in the case of soil erosion after fires, Shakesby (2011) found that at the plot level, many of the studied fires had erosion levels below 1 t ha^{-1}, with most below 10 t ha^{-1}. Therefore, it is necessary to consider the concept of an unnatural fire regime to explain the relationship between wildfires and desertification, also assuming that the degree of desertification will depend on soil type, climate, vegetation, and pre-fire degradation levels (Neary 2018). The current transformation of agricultural landscapes due to anthropogenic causes can lead to larger and more intense fires in a climate change context. In ecosystems sensitive to the effects of wildfires that lack adaptations to this type of disturbance, intense and severe fires can be a primary cause of desertification.

Box 6.1: Different Types of Effects and Elements Affected by Fire

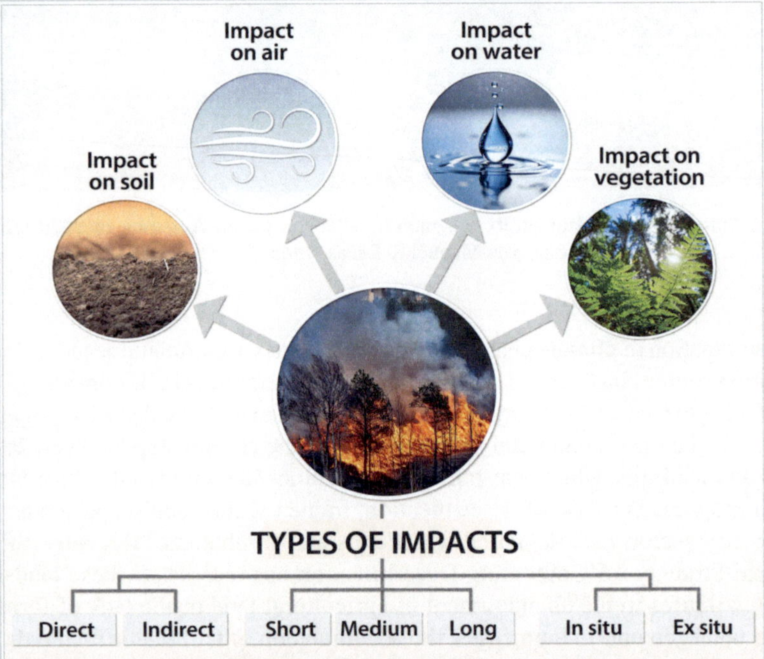

Wildfires impact soil, vegetation, air, and water, with effects that can be direct or indirect, short- or long-term, occurring both in situ (where the fire occurred) and at a distance. In soil, they promote erosion and degradation; in vegetation, they alter structure, composition, and ecological succession. Atmospherically, they release pollutants that affect air quality and public health. In water, ash, and sediments degrade quality and increase the risk of eutrophication. These impacts can accelerate desertification and reduce ecosystem recovery capacity, especially with recurrent fires and climate change.

Different Types of Impacts and Affected Elements

After a wildfire, the loss of vegetation cover and the protection it provides to the soil against precipitation increases wind or water erosion rates. The complete or partial combustion of vegetation reduces the surface area available for processes such as evaporation and transpiration, decreases the system's water retention capacity, and removes obstacles for surface runoff, thereby increasing the energy and transport capacity of the generated runoff. Additionally, the soil becomes more vulnerable to runoff and erosion processes, which can degrade its quality and functionality (Lucas-Borja et al. 2020; Carmona-Yáñez 2023). The combustion and reduction of soil organic matter due to fire can also affect its physicochemical and biological properties and, consequently, its functionality. However, the magnitude of these impacts, which could be associated with a loss of environmental and ecological quality in a given area due to fire, under a natural fire regime and over the long term, generally allows the ecosystem to respond positively and recover from the disturbance (Pausas 2024). In the short term, the environmental, ecological, and landscape impact may be significant, but over time, the system will regain its functionality and properties in Mediterranean environments under a natural fire regime (Fig. 6.3).

Conversely, under the parameters of an unnatural fire regime, such as fires with very high severity and intensity due to large vertical and horizontal biomass accumulation, fires occurring in atypical seasons for this type of disturbance mainly due to climate change, or fires with a very short recurrence due to increased ignition sources, or with a different extent and type of fire than that to which plant species are adapted, significant degradation processes can occur (Fig. 6.4).

In the context of an unnatural or anthropogenic fire regime, wildfires and desertification are interrelated in a feedback cycle that can have significant negative impacts on ecosystems, their properties, and functionality. Therefore, it is crucial to acknowledge that it is the fire regime, not the fire itself, that would trigger significant ecosystem degradation processes and consequently desertification. As demonstrated by Nolan et al. (2021), in an unnatural fire regime, changes in vegetation structure, demographics, and species composition are likely to occur (Fig. 6.4), with subsequent effects and degradation in ecosystems, potentially leading to significant desertification processes. In this context of an unnatural wildfire regime, this type of disturbance triggers an environmental and ecological degradation cycle that affects the health and stability of forest ecosystems and associated populations, as well as a loss of their capacity to recover and sustain biodiversity levels and ecosystem services that existed before the fire.

Fig. 6.3 Vegetation succession after fire. Sierra de los Donceles (Hellín, Castilla-La Mancha, Spain) fire in 2012. Photographs of the same areas in 2012 (**A**), 2014 (**B**), and 2023 (**C**). Photos by Manuel E. Lucas-Borja

Fig. 6.4 Landscape and vegetation composition changes in the 2009 Majadas fire (Serrania de Cuenca, Castilla La Mancha, Spain). Photos by Manuel E. Lucas-Borja

References

Carmona-Yáñez MD, Lucas-Borja ME, Zema DA et al (2023) Influence of management and stand composition on multifunctionality of Mediterranean tree forests. Trees 37:1801–1816. https://doi.org/10.1007/s00468-023-02462-w

Lucas-Borja ME, Delgado-Baquerizo M, Muñoz-Rojas M et al (2020) Changes in ecosystem properties after postfire management strategies in wildfire-affected Mediterranean forests. J Appl Ecol https://doi.org/10.1111/1365-2664.13819

Moreno M, Bertolín C, Arlanzón D et al (2023) Climate change, large fires, and cultural landscapes in the Mediterranean basin: an analysis in southern Spain. Heliyon 9(6):e16941. https://doi.org/10.1016/j.heliyon.2023.e16941

Neary DG (2018) Wildfire contribution to desertification at local, regional, and global scales. In: Squires VR, Ariapour A (eds) Desertification: past, current and future trends. Nova Science Publishers, Hauppauge, NY, pp 199–222

Nolan RH, Boer MM, Collins L et al (2021) Limits to post-fire vegetation recovery under climate change. Plant Cell Env 44:3471–3489. https://doi.org/10.1111/pce.14176

Pausas JG (2024) Incendios forestales. Una introducción a la ecología del fuego. Catarata y CSIC, Colección: ¿Qué sabemos de? núm. 158

Pausas JG, Fernández-Muñoz S (2011) Fire regime changes in the Western Mediterranean Basin: from fuel-limited to drought-driven fire regime. Clim Change 110:215–226. https://doi.org/10.1007/s10584-011-0060-6

Pausas JG, Llovet J, Rodrigo A et al (2008) Are wildfires a disaster in the Mediterranean basin? A review. Int J Wildland Fire 17:713–723. https://doi.org/10.1071/WF07151

Resco de Dios V (2020) Plant-fire interactions: applying ecophysiology to wildfire management. Springer Nature, Cham. https://doi.org/10.1007/978-3-030-41194-7

Sayedi SS, Abbott BW, Vannière B et al (2024) Assessing changes in global fire regimes. Fire Ecol 20:18. https://doi.org/10.1186/s42408-023-00237-9

Shakesby RA (2011) Post-wildfire soil erosion in the Mediterranean: review and future research directions. Earth Sci Rev 105:71–100. https://doi.org/10.1016/j.earscirev.2011.01.001

Turco M, Rosa-Cánovas JJ, Bedia J et al (2018) Exacerbated fires in Mediterranean Europe due to anthropogenic warming projected with non-stationary climate-fire models. Nat Commun 9:3821. https://doi.org/10.1038/s41467-018-06358-z

Chapter 7
Plant Invasions as a Symptom and Driver of Degradation

Ana Novoa and Pablo González-Moreno

Abstract Alien plant invasions are a crucial yet often underestimated component of desertification processes in Spain. Their ecological and socio-economic impacts are increasingly evident, involving both species with a long history of introduction (e.g., *Arundo donax*) and those more recently introduced (e.g., *Cenchrus setaceus*). For example, invasive alien plants threaten native biodiversity, reduce the productivity of agricultural and grazing lands, and alter key ecosystem processes such as water availability, soil stability, and fire regimes. In Spain's arid and semi-arid landscapes, where natural resilience is already compromised, the invasion of alien plants can accelerate land degradation and exacerbate desertification. Furthermore, their interaction with other global change factors can create feedback loops that further destabilize these fragile ecosystems. Integrating the management of invasive alien plants into broader environmental and restoration programs is essential to prevent further degradation and to maintain ecosystem services and supporting local communities dependent on these vulnerable drylands. Effective management can thus play a pivotal role in mitigating desertification and fostering long-term ecological sustainability in Spain.

A. Novoa (✉)
Estación Experimental de Zonas Áridas (EEZA), CSIC, Almería, Spain
e-mail: novoa@eeza.csic.es

Department of Invasion Ecology, Institute of Botany, Czech Academy of Sciences, Průhonice, Czech Republic

P. González Moreno
Research Group Evaluation and Restoration of Agroforest Systems, ERSAF, Department of Forestry Engineering, Universidad de Cordoba, Córdoba, Spain

ERSAF, Andalusian Institute for Earth System Research (IISTA), University of Córdoba, Córdoba, Spain
e-mail: ir2gomop@uco.es

J. Martínez-Valderrama and J. Olcina Cantos (eds.), *The Labyrinth of Desertification*,
https://doi.org/10.1007/978-3-032-09648-7_7

Keywords Alien plant invasions · Desertification · Environmental impacts · Invasive species management · Arid regions · Spain · Socio-economic impacts

Introduction to Alien Plant Invasion

For centuries, humans have introduced plant species to areas beyond their natural ranges, both intentionally (e.g., for ornamental, agricultural, or forestry purposes) and unintentionally (e.g., through the transport of goods, such as seeds in commercial soil; or via tourism, such as seeds attached to hiking boots; Hulme 2009). These species movements have intensified in recent decades due to globalization, and as a result, many plant species have reached new regions of the world where they did not previously occur (Seebens et al. 2021).

Only a portion of all introduced species establish self-sustaining populations in their new environments. Some also exhibit a strong capacity to spread, colonizing large areas and causing negative environmental and socio-economic impacts. The latter are known as invasive alien plants: a subset of introduced alien plants that became established and spread in the introduced regions, causing negative impacts on nature and often also on people (Roy et al. 2024).

It is estimated that 6% of all introduced plant species become invasive. The impacts of invasive plants include native biodiversity loss, alteration of biogeochemical cycles and ecological processes such as fire regimes or water cycles, as well as detrimental effects on economic sectors (e.g., on agriculture, tourism, or water management) and on human well-being (Roy et al. 2024). Furthermore, the management of alien plant invasions generate high economic costs for public administrations and society at large (Novoa et al. 2024).

In particularly fragile ecosystems, such as the arid and semi-arid environments of southern Europe, invasive alien plants pose an even greater threat, as they can undermine ecosystem resilience to other drivers of change traditionally associated with desertification processes. These ecosystems are already highly vulnerable to climate change, resource overexploitation, and land degradation. Plant invasions can amplify these pressures, acting synergistically with them and driving profound ecological transformations. In some cases, plant invasions may even act as triggers of desertification processes, accelerating the loss of key ecosystem services and threatening the ecological and socio-economic sustainability of affected regions (Ravi et al. 2009).

Plant Invasions and Desertification

Invasive alien plants can cause or accelerate the degradation of land in arid and semiarid ecosystems through various mechanisms. For example, some species have high transpiration rates and rapid growth, consuming large amounts of water. This is the case of the giant reed (*Arundo donax*), native to riparian zones in East Asia and invasive in Spain (Fig. 7.1), which can consume up to 2000 l of water per square meter. This reduces water availability for native species and can alter local hydrological regimes (Jiménez-Ruiz et al. 2021). In addition, giant reed invasions reduce the water flow capacity of riverbeds, causing major impacts (Box 7.1).

Fig. 7.1 Invasion of giant read (*Arundo donax*) in Cabo de Gata, Almería. Photo by Ana Novoa

Box 7.1: Media Interest in the Invasion of *Arundo donax* Following DANA Episodes in the Valencian Community

The invasion of the giant reed (*Arundo donax*) in riverbeds and agricultural fields across the Valencian Community has attracted the attention of various media outlets, especially after the episodes of heavy rainfall associated with the DANA in 2024. The invasion of the giant reed worsened the effects of the DANA-related flooding, negatively impacting local agriculture. Some examples of this media coverage include:

Una caña invasora arrastrada por la DANA coloniza el campo valenciano: "Superan la altura de árboles" (*An invasive reed dragged by the DANA colonizes the Valencian countryside: "They grow taller than trees")* – Europa Press, May 6, 2025.

Una especie invasora se dispara tras la DANA y amenaza el campo valenciano: "Se extiende como la pólvora" (*An invasive species surges after the DANA and threatens the Valencian countryside: "It spreads like wildfire")* – 20minutos, May 4, 2025.

La "Arundo donax," la caña invasora que agravó la riada de Valencia: "Quitar un kilómetro cuesta un millón" (*"Arundo donax," the invasive reed that worsened the Valencia flood: "Removing one kilometer costs a million")* – El Español, Novembeer 11, 2024.

Una caña invasora arrastrada por la DANA coloniza el campo valenciano (*An invasive reed dragged by the DANA colonizes the Valencian countryside*) – Telecinco, May 6, 2025.

These reports reflect growing social and political concern about the impact of the giant reed in the region, as well as the urgent need to implement management and control measures to mitigate its effects.

Invasive plants can also increase fuel biomass, leading to more frequent and intense wildfires (Brooks et al. 2004; Chap. 6). This shift in fire regimes can hinder ecosystem recovery by limiting secondary succession and accelerating land degradation. In Spain, several invasive plants, such as *Arundo donax* and *Ampelodesmos mauritanica*, contribute to changes in wildfire regimes by increasing both the quantity and continuity of fuel, both vertically and horizontally, thereby facilitating the rapid spread of intense wildfires (Vilà et al. 2001; Coffman et al. 2010).

Moreover, invasive alien plants often displace native species, reducing biodiversity and compromising the provision of key ecosystem services—such as soil stabilization and the maintenance of biogeochemical cycles—which in turn weakens ecosystem resilience to extreme events (Vilà et al. 2024). Invasive plants usually form dense and monospecific stands that limit light and water availability for native species, a particularly critical issue in arid and semiarid ecosystems. Notable examples include *Carpobrotus* spp. and *Arundo donax*, whose dominance can significantly alter native plant communities, and even affect other taxa, such as invertebrates (Campoy et al. 2018; Jiménez-Ruiz et al. 2021).

Alien plant invasions also pose a significant threat to human well-being. On one hand, they cause exorbitant economic losses. For example, in Spain, the cost of alien plant invasions has amounted to at least over 140 million euros since 1997 (Novoa et al. 2024). The costliest species is the water hyacinth (*Eichhornia crassipes*), an aquatic plant that invades lakes and rivers, producing large amounts of biomass. Its rapid proliferation drastically reduces the amount of light penetrating the water and limits gas exchange between the atmosphere and the aquatic environment. In addition, due to its high evapotranspiration rate, it decreases the availability of water, which affects supply for irrigation, human consumption, and other uses. The water hyacinth also clogs navigation and irrigation channels, hinders economic and recreational activities, and creates ideal conditions for mosquito proliferation. Currently, the recorded costs of water hyacinth in Spain exceed 53 million euros (Novoa et al. 2024). In other parts of the world, such as Lake Victoria (Mujingni Epse Cho 2012), the invasion of water hyacinth has even forced communities to abandon their livelihoods.

Other examples of alien invasive plants causing major economic losses in Spain include eucalyptus (*Eucalyptus* spp.), with losses of up to 41 million euros, the giant reed (*Arundo donax*), with losses of over 11 million euros, and fountaingrass (*Cenchrus setaceus*, Fig. 7.2), with losses exceeding 8 million euros (Novoa et al. 2024).

In arid regions, certain types of exotic plants tend to show a higher propensity to invade than others. For instance, it has been observed that a large number of plants exhibiting Crassulacean Acid Metabolism (CAM), once introduced into arid areas outside their native range, have managed to spread and invade (Yu et al. 2024). This is likely due to their physiological adaptations, which give them advantages in environments characterized by water scarcity, high temperatures, and poor soils—conditions that may facilitate their success as invasive species in arid ecosystems. For example, several species from the cactus family (Cactaceae), especially those in the *Opuntia*, *Austrocylindropuntia,* and *Cylindropuntia* genera, are invading arid areas of Africa, Australia, the United States, and Europe (Novoa et al. 2015, Fig. 7.3), causing a wide range of negative ecological and socioeconomic impacts (Novoa et al. 2016).

Alien plant invasions interact with other global change drivers such as climate change, overexploitation of natural resources, land-use changes, and pollution, creating complex scenarios that worsen the loss of ecosystem functionality and resilience (Roy et al. 2024). For example, the overexploitation of aquifers reduces the resilience of native vegetation, facilitating invasions by alien plant species that access groundwater more efficiently (Dudley et al. 2014). Similarly, the combination of recurring fires and prolonged droughts creates a hostile environment for native species, favoring the invasion of fire-adapted alien species (Le Maitre et al. 2023). These interactions generate feedback loops that can accelerate the processes of invasion and desertification (Ravi et al. 2009).

Fig. 7.2 Invasion of crimson fountaingrass (*Cenchrus setaceus*) in Córdoba, Southern Spain. Photos by Pablo González

Fig. 7.3 Invasion of *Austrocylindropuntia subulata* in Cabo de Gata, Almería. Photo by Ana Novoa

Addressing Plant Invasions

The most cost-effective way to address alien plant invasions is to prevent the introduction of new potentially invasive alien plants (Roy et al. 2024). This requires predicting which species could become invasive if introduced, and regulating their introduction. It is also essential to have clear and effective protocols for the early detection of and rapid response to new invasions.

Since 2007, Spain has had a specific legal framework to address invasive alien species through the Spanish Catalog of Invasive Exotic Species. Established under the Law on Natural Heritage and Biodiversity and developed through Royal Decree 630/2013, the catalog prohibits the possession, transport, and trade of listed species, and mandates their control or eradication. Inclusion in the catalog is based on scientific evidence regarding human-mediated introduction and the species' potential threat to biodiversity, agriculture, or the economy. Currently, the catalog includes 199 taxa, 85 of which are plants.

Despite these legal restrictions, enforcement remains challenging, particularly in private land, which undermines the long-term effectiveness of control and restoration efforts. In addition to the catalog, a list of 3,500 potentially invasive alien taxa hosted on the website of MITECO (Ministry for the Ecological Transition and the Demographic Challenge). Royal Decree 570/2020 (16 June) establishes that importing any of these taxa from outside the European Union for the first time requires prior environmental authorization from MITECO.

Finally, since 2014, there has been an EU regulation (No. 1143/2014, on the prevention and management of the introduction and spread of invasive alien species), which includes a list of invasive alien species of Union concern and introduces restrictions similar to those in the Spanish legislation (Castro Díez et al. in press).

Acknowledgments AN was supported by the MCIN/AEI/10.13039/501100011033 and the FSE + (grant No. RYC2022-037905-I) and by long-term research development project RVO 67985939 (Czech Academy of Sciences). PGM was supported by the RyC fellowship (RYC2021-033138-I), funded by MCIN/AEI/https://doi.org/10.13039/501100011033 and European Union ("NextGenerationEU"/PRTR).

References

Brooks ML, D'Antonio CM, Richardson DM, Grace JB, Keeley JE, DiTomaso JM, Hobbs RJ, Pellant M, Pyke D (2004) Effects of invasive alien plants on fire regimes. BioScience 54(7):677–688

Campoy JG, Acosta AT, Affre L, Barreiro R, Brundu G, Buisson E, González L, Lema M, Novoa A, Retuerto R, Roiloa SR, Fagúndez J (2018) Monographs of invasive plants in Europe: Carpobrotus. Botany Lett 165(3-4):440–475

Castro Díez P, Alonso A, Gallardo B, Vilà M, García-Berthou E, Novoa A, Angulo E, Calmaestra RG (in Press) Diez años de avances en investigación y gestión de invasiones biológicas en España. Ecosistemas

Coffman GC, Ambrose RF, Rundel PW (2010) Wildfire promotes dominance of invasive giant reed (*Arundo donax*) in riparian ecosystems. Biol Invasions 12(8):2723–2734

Dudley BD, Flint Hughes R, Ostertag R (2014) Groundwater availability mediates the ecosystem effects of an invasion of *Prosopis pallida*. Ecol Appl 24(8):1954–1971

Hulme PE (2009) Trade, transport and trouble: managing invasive species pathways in an era of globalization. J Appl Ecol 46(1):10–18

Jiménez-Ruiz J, Hardion L, Del Monte JP, Vila B, Santín-Montanyá MI (2021) Monographs on invasive plants in Europe N° 4: *Arundo donax* L. Botany Lett 168(1):131–151

Le Maitre DC, Máguas C, Ulm F, Marchante H (2023) Ecological impacts and changes in ecosystem services and disservices mediated by invasive Australian Acacia species. In: Richardson DM, Le Roux JJ, Marchante E (eds) Wattles: Australian Acacia Species around the World. CABI, Wallingford, UK

Mujingni Epse Cho JT (2012) Quantification of the impacts of water hyacinth on riparian communities in Cameroon and assessment of an appropriate method of control: the case of the Wouri River Basin. World Maritime University Dissertations, Malmö, Sweden. http://commons.wmu.se/all_dissertations/29

Novoa A, Kumschick S, Richardson DM, Rouget M, Wilson JR (2016) Native range size and growth form in Cactaceae predict invasiveness and impact. NeoBiota 30:75–90

Novoa A, Le Roux JJ, Robertson MP, Wilson JR, Richardson DM (2015) Introduced and invasive cactus species: a global review. AoB Plants 7:plu078

Novoa A, Oficialdegui FJ, Gómez-Calmaestra R, Durán C, Angulo E (2024) ¿Cuánto cuestan las invasiones biológicas al bolsillo de los españoles?. Quercus 458:20–27

Ravi S, D'Odorico P, Collins SL, Huxman TE (2009) Can biological invasions induce desertification? New Phytol 181:512–515

Roy HE, Pauchard A, Stoett PJ et al. (2024) Curbing the major and growing threats from invasive alien species is urgent and achievable. Nat Ecol Evol 8(7):1216–1223

Seebens H, Bacher S, Blackburn TM, Capinha C, Dawson W, Dullinger S, Genovesi P, Hulme PE, Van Kleunen M, Kühn I, Jeschke JM (2021) Projecting the continental accumulation of alien species through to 2050. Global Change Biol 27(5):970–982

Vilà M, Trillo A, Castro-Díez P, Gallardo B, Bacher S (2024) Field studies of the ecological impacts of invasive plants in Europe. NeoBiota 90:139–159

Vilà M, Lloret F, Ogheri E, Terradas J (2001) Positive fire–grass feedback in Mediterranean Basin woodlands. Forest Ecol Manag 147(1):3–14

Yu K, D'Odorico P, Novoa A, Collins SL, Hartzell S, Huang H, Liu H, Weigelt P, Porporato AM (2024) Potential expansion of plants with crassulacean acid metabolism in the Anthropocene. BioScience 74(7):478–487

Chapter 8
Forest Mortality: How Tree Death Can Accelerate Desertification

Rafael Ma Navarro-Cerrillo

Abstract Tree mortality has long-term ecological effects on the structure and functioning of ecosystems. One concern arising from increased forest mortality is the potential for aggravating desertification processes. The rapid and extensive reduction in vegetation cover can modify eco-hydrological processes, causing greater net losses of water and soil that feed back into the loss of productivity and vegetation cover vigor, generating a spiral of desertification. Once these processes (mortality-desertification) begin, changes in vegetation loss and erosion patterns can persist for decades. Preventing forest mortality through forest management may not be feasible due to the large extent of the potentially affected areas. Density regulation (e.g., thinning) can be a management strategy to reduce the vulnerability of forest stands to mortality processes, and thus desertification. Although in general terms, the limited management options to prevent forest mortality highlight the need for new research to improve the ability to predict the ecohydrological consequences of mortality and its impact on desertification.

Keywords Tree mortality · Desertification · Ecohydrology · Vegetation cover · Forest management

R. M. Navarro-Cerrillo (✉)
Department of Forest Engineering. Dendrochronology, Silviculture, and Climate Change Laboratory, DendrodatLab-ERSAF, University of Córdoba, Córdoba, Spain
e-mail: rmnavarro@uco.es

J. Martínez-Valderrama and J. Olcina Cantos (eds.), *The Labyrinth of Desertification*,
https://doi.org/10.1007/978-3-032-09648-7_8

Mortality Processes in Forest Systems

Ecosystem dynamics are altered by disturbances, whether natural or anthropogenic (McDowell et al. 2020). In the case of forest mortality events, environmental stress can directly cause tree death, which accumulates linearly at broader spatial scales—individual, stand, and forest—or interact with other disturbances to propagate nonlinearly at the landscape or even regional scales (Fig. 8.1). Due to these interactions across spatial scales, different explanations have been offered for forest mortality (Hartmann et al. 2022). Recently, the combination of extreme droughts and elevated temperatures has apparently surpassed the physiological mortality thresholds of multiple tree species (Sala et al. 2010). However, although direct climate-induced stress may have been sufficient to cause mass mortality in some forest systems, important biotic agents affecting weak and dying trees have been identified, obscuring the underlying and primary causes of widespread forest mortality patterns (Hartmann et al. 2022). Furthermore, there is considerable uncertainty about how the variability of climate factors will increase, resulting in increased frequency, severity (both temporal and spatial), and severity of droughts.

These mass mortality of a dominant species can have long-term ecological effects and could have profound consequences for ecosystem structure and functioning, species composition, ecohydrological processes, biogeochemical cycling, disturbance dynamics, and the provision of ecosystem services, including carbon sequestration (Anderegg et al. 2015). The recent mortality processes in the southeast of the Iberian Peninsula (Fig. 8.1) perfectly illustrate these impacts at different spatial scales, driven by positive feedback between water stress and biotic agents (Trujillo-Toro and Navarro-Cerrillo 2019). One concern arising from the possible increase in forest mortality is the possibility that it will aggravate desertification processes; Particularly in systems that are already highly vulnerable to these processes, such as Mediterranean forest ecosystems, given their climatic limitations, the dominance of systems at the limits of the distribution of many tree species, and potential changes in forest structure, with high levels of bare soil and low herbaceous cover (see the case studies "Shrub encroachment in drylands" and "Forest fires and desertification: The role of the fire regime").

Effect of Mortality Processes on Desertification

The rapid and extensive reduction of vegetation cover—not only woody but also herbaceous—below critical threshold values due to tree death, which directly affects the structure and spatial distribution of vegetation, can modify both eco-hydrological processes and desertification (Adams et al. 2012). When analyzing mortality processes at the landscape scale, they cause direct eco-hydrological effects, specifically on the components of the water balance (soil evaporation-E-, plant transpiration -T-, evapotranspiration -ET-, and evaporation of water intercepted by the plant canopy -I-) (del Campo et al. 2014; Fig. 8.2).

Fig. 8.1 Mass mortality processes of *Pinus halepensis* in Guadix (Granada, southern Spain) operating at the individual tree (**A**), stand (**B**), and landscape (**C**) scales. Photos by Rafael Navarro-Cerrillo

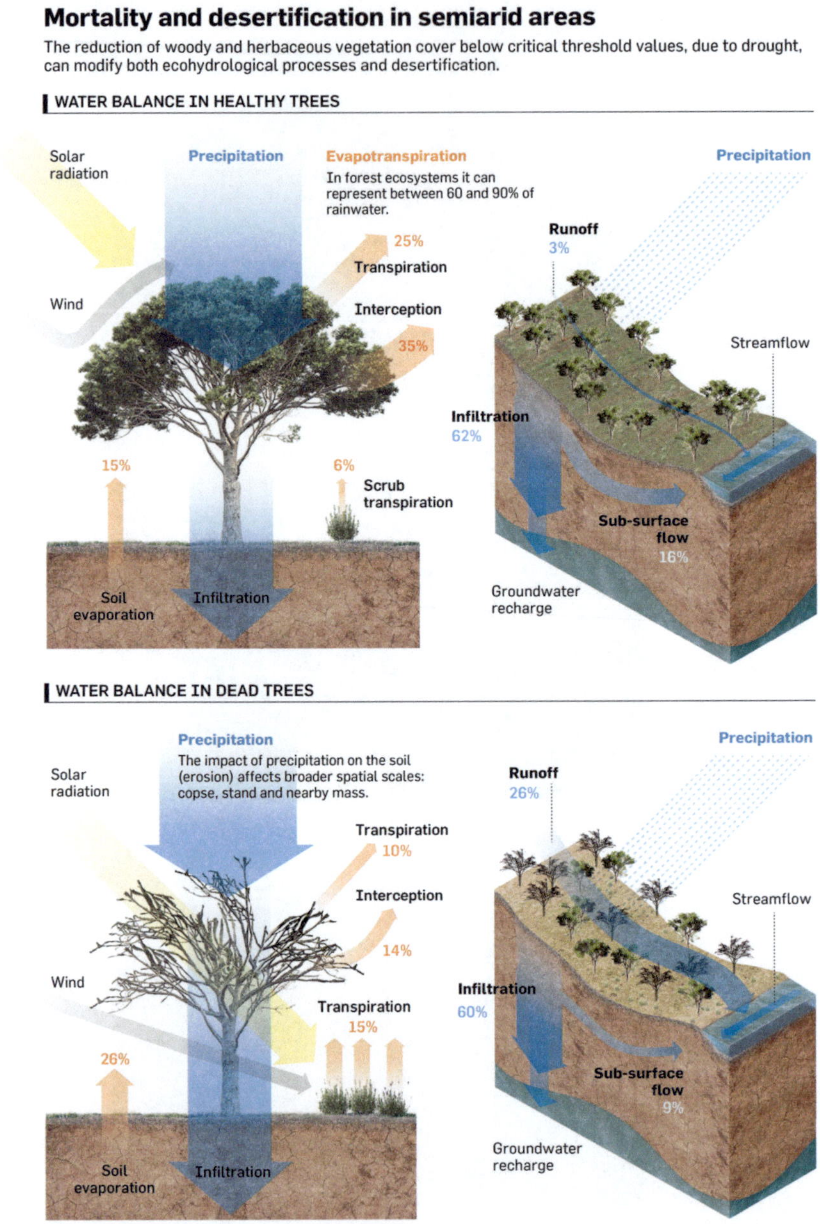

Fig. 8.2 Diagram representing the interactions across spatial scales between forest mortality and desertification in semiarid areas of the southeastern Iberian Peninsula. Based on Adams et al. (2012) and del Campo et al. (2014)

Considering these changes in the water balance, mortality can indirectly affect other aspects of hydrological functioning, such as infiltration and flow path distribution, which would subsequently result in changes in soil moisture status, groundwater recharge dynamics, and flow volume and frequency. Furthermore, tree loss increases the "energy" reaching the soil. As a result, the reduction in canopy cover associated with forest mortality can lead to earlier, faster, and more synchronized soil drying, amplifying runoff and erosion as isolated tesserae of bare soil connect, merging into percolation networks at broader spatial scales (Adams et al. 2012). Greater net water and soil losses feedback into the loss of productivity and tree vigor to vegetation cover, generating a desertification spiral (Fig. 8.2). Once these processes (mortality-desertification) have begun, changes in the relationships between vegetation loss patterns and accelerated erosion processes can persist for decades, and, once the desertification process is consolidated by vegetation and soil loss, semiarid ecosystems can take decades to recover.

Studies that have addressed the effect of canopy cover reductions, due to factors other than mortality, on the hydrological balance (see, e.g., Molina et al. 2021) show the potentially opposing influences of the E and T components with different thinning weights (similar to cover loss). Intense density loss showed significant differences in soil water content, although transpiration in plots with severe thinning was remarkably similar to that in moderate thinning. These results support the idea that excessive forest openness causes the understory to compete more strongly for water, thus counteracting the greater tree transpiration observed in the short term.

In this sense, if mortality occurs in areas with low aridity index values, then forest mortality may lead to greater water availability for other vegetation components, as an indirect effect of tree mortality, although this will be qualified by the infiltration capacity of the soils. However, in semiarid watersheds, when tree cover is lost, a sudden reduction in post-mortality water yield can occur, either due to rapid understory growth, which reduces surface flow, or due to increase of E-T-I. Under severe drought conditions and severe forest mortality, deep soil water resources can be rapidly depleted, making vegetation whose roots extract water at shallower depths (e.g., plantations of *Pinus pinaster*) highly susceptible. This can lead to potential nonlinear thresholds between mortality and hydrological response, and to specific effects on runoff-generating processes (i.e., *Hortonian* overland flow, saturation excess overland flow, and subsurface flow).

Management Alternatives to Reduce the Impact of Mortality Processes on Desertification

Forest mortality in areas highly vulnerable to desertification (low aridity index, high drought frequency) is beginning to be noticeable, and significant changes in vegetation cover have already occurred, mainly due to emergency forestry interventions (e.g., Iberian southern mountains, Navarro-Cerrillo et al. 2019), similar to those experienced after major fires. It would be important to explore these changes in greater depth by integrating field hydrological and remote sensing data to assess changes in woody cover. Patterns at different scales may show the loss of tree vegetation, but these may obscure other, more subtle changes, such as a rapid increase in shrub cover. This demonstrates the need for more detailed and dynamic analyses to address temporal trends in mortality processes at the local scale (Ariza Salamanca et al. 2019, Fig. 8.3). The end result can lead to more fragmented landscapes due to the recurrent action of severe drought periods that cause stand dieback along with the expansion of other vegetation structures (e.g., shrublands), as has already been observed in areas affected by severe mortality processes (e.g., Guadix; Fig. 8.1). This situation means that management alternatives following a forest mortality process will likely differ depending on the climate and type of system affected, as well as the specific objectives of the affected area (Navarro-Cerrillo et al. 2019). Preventing large-scale forest mortality through forest management may not be feasible due to the significant extent of the potentially affected areas, which makes the planning and cost of management extremely difficult. Furthermore, the current state of knowledge regarding the impact of forest mortality on desertification prevents an accurate prediction of the effects in a specific watershed.

Studies on the effect of forest management, through improvement of silvicultural practices (e.g., thinning), suggest that these can be a management strategy to reduce the vulnerability of forest stands to drought and pests (Sohn et al. 2016). Several studies have shown that trees reduce their vulnerability to drought after heavy or moderate

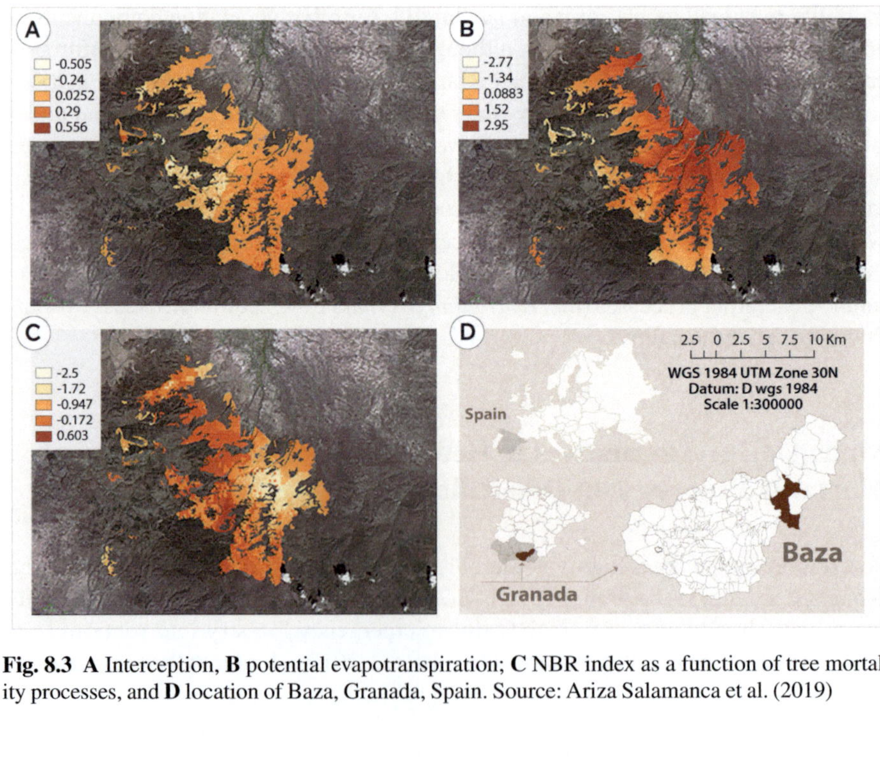

Fig. 8.3 **A** Interception, **B** potential evapotranspiration; **C** NBR index as a function of tree mortality processes, and **D** location of Baza, Granada, Spain. Source: Ariza Salamanca et al. (2019)

thinning (Del Campo et al. 2022; Navarro-Cerrillo et al. 2023), and that this effect persists over time, although these responses depend on the species (Manrique-Alba et al. 2020). Post-mortality silvicultural options to mitigate potential desertification processes probably consist of promoting desirable species better adapted to the new conditions including regeneration and structural diversification. However, many forests show extremely limited regeneration due to poor adaptation to current climatic conditions, which converts many of the stands affected by forest mortality into "finalist stands" (not subject to regeneration) (Navarro-Cerrillo et al. 2019). Overall, the limited management options for preventing forest mortality over large areas highlight the need for new research to improve our ability to predict the ecohydrological consequences of mortality processes and their impact on desertification.

References

Adams HD, Luce CH, Breshears DD, Allen CD, Weiler M, Hale VC, Huxman TE (2012) Ecohydrological consequences of drought-and infestation-triggered tree die-off: insights and hypotheses. Ecohydrol 5(2):145–159

Anderegg WR, Hicke JA, Fisher RA, Allen CD, Aukema J, Bentz B, Zeppel M (2015) Tree mortality from drought, insects, and their interactions in a changing climate. N Phytol 208(3):674–683

Ariza Salamanca AJ, Navarro-Cerrillo RM, Bonet-García FJ, Pérez-Palazón MJ, Polo MJ (2019) Integration of a Landsat time-series of NBR and hydrological modeling to assess *Pinus pinaster* Aiton. Forest defoliation in South-Eastern Spain. Remote Sens 11(19):2291

del Campo AD, Fernandes TJ, Molina AJ (2014) Hydrology-oriented (adaptive) silviculture in a semi-arid pine plantation: how much can be modified the water cycle through forest management?. Eur J Res 133:879–894

del Campo AD, Otsuki K, Serengil Y, Blanco JA, Yousefpour R, Wei X (2022) A global synthesis on the effects of thinning on hydrological processes: implications for forest management. For Ecol Manag 519:120324

Hartmann H, Bastos A, Das AJ, Esquivel-Muelbert A, Hammond WM, Martínez-Vilalta J, Allen CD (2022) Climate change risks to global forest health: emergence of unexpected events of elevated tree mortality worldwide. Annu Rev Plant Biol 73(1):673–702

Manrique-Alba À, Beguería S, Molina AJ, González-Sanchis M, Tomàs-Burguera M, del Campo AD, Camarero JJ (2020) Long-term thinning effects on tree growth, drought response and water use efficiency at two Aleppo pine plantations in Spain. Sci Total Env 728:138536

McDowell NG, Allen CD, Anderson-Teixeira K, Aukema BH, Bond-Lamberty B, Chini L, Xu C (2020) Pervasive shifts in forest dynamics in a changing world. Science 368(6494): eaaz9463.

Molina AJ, González-Sanchis M, Biel C, del Campo AD (2021) Ecohydrological turnover in overstocked Aleppo pine plantations: does the effect of thinning, in relation to water, persist at the mid-term?. For Ecol Manag 483:118781

Navarro-Cerrillo RM, Cachinero-Vivar AM, Pérez-Priego Ó, Cantón RA, Begueria S, Camarero JJ (2023) Developing alternatives to adaptive silviculture: thinning and tree growth resistance to drought in a *Pinus* species on an elevated gradient in Southern Spain. For Ecol Manag 537:120936

Navarro-Cerrillo RM, Sánchez-Salguero R, Rodriguez C, Lazo JD, Moreno-Rojas JM, Palacios-Rodriguez G, Camarero JJ (2019) Is thinning an alternative when trees could die in response to drought? The case of planted Pinus nigra and *P. sylvestris* stands in southern Spain. For Ecol Manag 433:313–324

Sala A, Piper F, Hoch G (2010) Physiological mechanisms of drought-induced tree mortality are far from being resolved. N Phytol 186(2):274–281

Sohn JA, Saha S, Bauhus J (2016) Potential of forest thinning to mitigate drought stress: a meta-analysis. For Ecol Manag 380:261–273

Trujillo-Toro J, Navarro-Cerrillo RM (2019) Analysis of site-dependent *Pinus halepensis* Mill. Defoliation caused by 'Candidatus *Phytoplasma pini*' through shape selection in Landsat time series. Remote Sens 11(16):1868

Chapter 9
Moving Livestock to Reverse Desertification Landscapes

Pablo Manzano

Abstract Grazing livestock is often identified as a vector for environmental degradation and desertification. Such consideration, arising from popular culture, ignores the vast natural extent of ecosystems with a limited tree cover, as well as the role of herbivore disturbances in maintaining them. Herbivore mobility is a fundamental precondition for the maintenance of natural ecological processes, and traditional livestock keeping has mimicked such mobility in order to sustainably produce food in grass-dominated ecosystems. While livestock mobility is crucial for sustainability, the complexity of ecosystems—and of the livestock social-ecological systems that interact with them—means that mobility alone is insufficient to guarantee ecological integrity. The socio-cultural and economic contexts that surround it, including elaborate practices relying on Traditional Ecological Knowledge but also local governance regulations, contribute to maintain ecological integrity. Erosion of the traditional grazing socio-ecosystem through fragmentation, but also loss of day-to-day pastoralist practices, market pressures from more high-yield livestock systems, and drives toward reduced manpower, evidence the threat that sustainable livestock systems are under. The whole socio-ecosystem of mobile pastoralism, including livestock mobility but also its surrounding cultural practices, must be safeguarded to ensure sustainable practice.

Keywords Herbivore ecology · Open ecosystems · Mobile pastoralism · Traditional ecological knowledge · Traditional governance · Agricultural intensification

P. Manzano (✉)
BC3 – Basque Centre for Climate Change, Leioa, Spain
e-mail: pablo.manzano@bc3research.org

Ikerbasque – Basque Foundation for Science, Bilbao, Spain

J. Martínez-Valderrama and J. Olcina Cantos (eds.), *The Labyrinth of Desertification*,
https://doi.org/10.1007/978-3-032-09648-7_9

Does Livestock Really Cause Desertification?

Livestock has conventionally been portrayed as a driver of desertification. However, such assumptions are based on perspectives that overlook the territorial complexity and recent advances in terrestrial ecosystem science. Early biogeographical paradigms, rooted in the work of Humboldt and Clemens, promoted the idea of uniformly forested landscapes with closed tree canopies. This view has since evolved into a broader recognition of the natural character of complex landscapes with extensive open areas (Bond 2019). These landscapes are sustained through frequent disturbances that remineralize much of the plant biomass, either via digestion by large herbivores or fire, effectively resetting the system (Pausas and Bond 2020).

Following the near-complete disappearance of large wild herbivores—particularly grazers—from continental ecosystems in the late Pleistocene (Fricke et al. 2022), extensive livestock grazing has replaced both the biomass of this functional guild (Greenspoon et al. 2023) and many of its ecological functions (Manzano et al. 2021). From a "land sparing" perspective—that is, ecosystem conservation through total exclusion of human activity—this substitution has come at the cost of an unsustainable environmental footprint (Poore and Nemecek 2018). However, this viewpoint fails to recognize the vast heterogeneity in environmental impacts across different types of livestock systems (Manzano et al. 2025). Even within extensive livestock systems, various management regimes lead to markedly different outcomes, particularly in relation to livestock mobility and the governance practices that underpin it (Manzano et al. 2023).

Box 9.1: Land Sparing-Sharing Paradigm

The land sparing-sharing paradigm contrasts two approaches to land management for biodiversity conservation and agricultural production. Land sparing involves high-yield agriculture on a smaller area, leaving larger tracts of land untouched for conservation. Land sharing integrates biodiversity conservation within agricultural landscapes, often through lower-yield, more environmentally friendly farming practices.

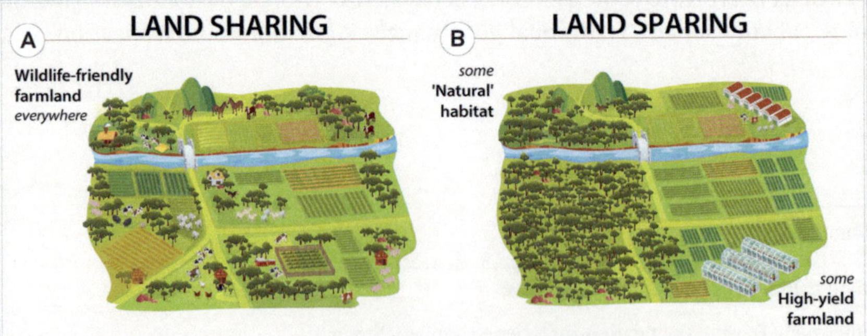

A Land Sharing:

Lower-yield agriculture: Emphasizes agricultural practices that are less intensive and more compatible with biodiversity at the expense of lower yield, and consequently, more land is utilized for food production.

Integration: Biodiversity is maintained within a favorable agricultural landscape, mitigating fragmentation effects.

Multiple goals: Aims to achieve both agricultural production and biodiversity conservation within the same area.

Example: Agroforestry systems where trees and crops are grown together, or organic farming with practices that limit pesticides and benefit wildlife.

B Land Sparing:

High-yield agriculture: Focuses on maximizing crop production on a smaller land area. Extensive use of pesticides and fertilizers. Dominance of linear nutrient flows that will pollute water bodies and have other environmental impacts.

Spatial separation: Keeps agricultural land distinct from natural habitats.

Conservation focus: Larger areas are set aside for pristine ecosystems, with the aim to protect biodiversity and other ecosystem services.

Example: Intensively farmed land with high-yield crops next to a protected forest or wilderness area.

Mobile Livestock as a Proxy of Naturally Migrating Grazing Herbivores

The choice of domesticating mainly grazer species (Hofmann 1989) during the Neolithic is not likely to be a random choice. In natural systems that have conserved complete herbivore guilds that execute migrations, such as the Greater Serengeti ecosystem, those are dominated in biomass by grazer species (Sinclair et al. 2009) that, due to their dietary adaptation, cannot use the standing biomass of woody plants during the dry season—being forced instead to follow the productivity peak of grasses. On the other hand, they avoid top-down regulation by carnivores that establish territories, a predator control that their grazer counterparts can't escape. This translates into large numbers and biomasses mostly controlled bottom-up by plant food availability (Mduma et al. 1999). Humans, with the aim of capturing the maximum possible amount of productivity and biomass in grassy ecosystem, would have concentrated in domesticating mostly grazing species such as cows, horses or sheep, but they would have had to keep mobility to sustain high productivity levels (Fig. 9.1).

Mobility can explain many of the positive environmental outcomes associated with more sustainable livestock systems, to the extent that it may have delayed desertification

Fig. 9.1 Recently shorn sheep grazing in an arid summer landscape. Photo by Elsa Varela

in the Sahara by as much as 500 years (Brierley et al. 2018). Its alignment with vegetation productivity, following the maturation cycles of grasslands (Manzano Baena and Casas 2010), facilitates a coupling between productivity and disturbance that has been proposed as a mechanism underlying high ecological diversity and functionality at global scale (Huston 1979, 1994). Measurable net effects observed in Spanish transhumant systems (Fig. 9.2) include improved recruitment of tree saplings—thanks to resting of pastures during critical establishment phases (Carmona et al. 2013)—as well as optimized seed dispersal by grazing animals during peak ripeness, and preservation of flowers that serve as pollinator forage, enhancing both fertility and reproductive capacity in plant communities (García-Fernández et al. 2019). Other traditional practices embedded in mobile systems, such as night penning or kraaling—widespread among many pastoralist societies—are known to contribute to the maintenance of high-productivity, resilient grassland patches (Augustine et al. 2003).

Box 9.2: "Green Deserts" Due to Undergrazing

Some authors (e.g., Perevolotsky and Seligman 1998) refer to grasslands that have been invaded by low-growing woody vegetation—shrubs—as "green deserts" due to the absence of grazing, which causes the land to lose economic value (and one meaning of desertification is the consideration of degradation as an economic loss). However, many ecologists will agree that such an increase in plant biomass is a clear indication of ecosystem recovery (e.g., Maestre et al. 2009), as do many of the indicators used to monitor the advance or retreat of desertification.

Shrub-invaded grassland due to lack of grazing. Photo by Elsa Varela

Disruptions in Conserved Mobile Systems Driving to Environmental Degradation

Nonetheless, transhumant and nomadic systems, even where they have survived to the present, are not immune to sustainability challenges associated with global change. The decline of extensive grazing operations and increase in average herd size (Delattre et al. 2020; Galán et al. 2022)—a necessary adaptation to compete with more productive industrial systems (Manzano et al. 2025)—has led to the abandonment of environmentally beneficial practices in summer and winter pastures, such as adaptive rotational grazing and kraaling, both of which are labor-intensive. Loss of soil fertility encourages the dominance of shrub species (Pauné Fabré 2021), which—as discussed in Chap. 4 of this volume—can trigger processes of desertification. Furthermore, imbalances in the grazer-to-browser ratio within herbivore communities may lead to shrub encroachment, ecosystem degradation, and increased fire risk, both due to fuel continuity and decreased biodiversity (Velamazán et al. 2023; Serrano-Zulueta et al. 2023). These changes may be the result of livestock mobility decline, but also of reduced stocking rates driven by economic factors: as the primary sector weakens, pasture becomes more abundant, allowing animals to be more selective in their foraging (Pauné Fabré 2021). The lack of manpower is also likely to reduce bunch grazing and allow grazing animals to select their food, which will decrease palatable species and

Fig. 9.2 **A** Sheep grazing in the undergrowth of a Mediterranean pine forest. While feeding, the animals help reduce fire risk by clearing vegetation and convert plant fibers—digestible only by ruminants—into protein. Photo by Elsa Varela. **B** Cattle grazing on high mountain pastures in the Sierra de Baza (Granada, Spain), located in the interior of the southeastern Iberian Peninsula. Photo by Jaime Martínez Valderrama

increase unpalatable ones. While this is a well-known effect in livestock science, the disruption of natural processes it constitutes has been understood through the formulation of landscapes of fear, i.e., the presence of predators forcing wild herbivores to bunch together and graze in a non-selective manner (Riginos 2015) (Fig. 9.2).

Traditional ecological knowledge—such as the timing of transhumant movements—is also eroding, leading to a decoupling between domestic herbivory and natural processes, now further disrupted by climate change (Manzano 2024), with potentially adverse effects for ecosystem maintenance. Differences in the dates when pasture is grazed between long- and short-distance transhumance in the Sierra de Segura in Spain, also show the effects that such mismatches can have. Long-distance transhumance is able to save pastures from early grazing because of making use of available grass in the drove roads. The early arrival of herds in the short-distance grazing system disbalances the plant community, disproportionally affecting perennials and favoring annuals (Parra et al. 2025). This not only has consequences in terms of community composition, but also in the productivity of pastures and hence ability to provide food services. Perennial species are able to increase faster the grass biomass at the beginning of the growing season because they depart from a large pre-existing biomass of photosynthetic tissue, which annuals have to generate from seeds in the seed bank.

References

Augustine DJ, McNaughton SJ, Frank DA (2003) Feedbacks between soil nutrients and large herbivores in a managed savanna ecosystem. Ecol Appl 13:1325–1337. https://doi.org/10.1890/02-5283

Bond WJ (2019) Open ecosystems: ecology and evolution beyond the forest edge. Oxford University Press, Oxford. https://doi.org/10.1093/oso/9780198812456.001.0001
Brierley C, Manning K, Maslin M (2018) Pastoralism may have delayed the end of the green Sahara. Nat Commun 9:4018. https://doi.org/10.1038/s41467-018-06321-y
Carmona CP, Azcárate FM, Oteros-Rozas E, González JA, Peco B (2013) Assessing the effects of seasonal grazing on holm oak regeneration: implications for the conservation of Mediterranean dehesas. Biol Conserv 159:240–247
Delattre L, Debolini M, Paoli JC, Napoleone C, Moulery M, Leonelli L, Santucci P (2020) Understanding the relationships between extensive livestock systems, land-cover changes, and CAP support in less-favored Mediterranean areas. Land 9:518
Fricke EC, Hsieh C, Middleton O, Gorczynski D, Cappello CD, Sanisidro O, Rowan J, Svenning J-C, Beaudrot L (2022) Collapse of terrestrial mammal food webs since the Late Pleistocene. Science 377:1008–1011. https://doi.org/10.1126/science.abn4012
Galán E, Garmendia E, García O (2022) The contribution of the commons to the persistence of mountain grazing systems under the Common Agricultural Policy. Land Use Policy 117:106089. https://doi.org/10.1016/j.landusepol.2022.106089
García-Fernández A, Manzano P, Seoane J, Azcárate FM, Iriondo JM, Peco B (2019) Herbivore corridors sustain genetic footprint in plant populations: a case for Spanish drove roads. PeerJ 7:e7311. https://doi.org/10.7717/peerj.7311
Greenspoon L, Krieger E, Sender R, Rosenberg Y, Bar-On YM, Moran U, Antman T, Meiri S, Roll U, Noor E, Milo R (2023) The global biomass of wild mammals. Proc Natl Acad Sci USA 120:e2204892120. https://doi.org/10.1073/pnas.2204892120
Hofmann RR (1989) Evolutionary steps of ecophysiological adaptation and diversification of ruminants: a comparative view of their digestive system. Oecologia 78:443–457. https://doi.org/10.1007/BF00378733
Huston MA (1979) A general hypothesis of species diversity. Am Nat 113:81–101
Huston MA (1994) Biological diversity: the coexistence of species. Cambridge University Press, Cambridge, MA
Maestre FT, Bowker MA, Puche MDM, Hinojosa B, Martínez I, García-Palacios P, Castillo AP, Soliveres S, Luzuriaga AL, Sánchez AM, Carreira JA, Gallardo A, Escudero A (2009) Shrub encroachment can reverse desertification in semi-arid Mediterranean grasslands. Ecol Lett 12:930–941. https://doi.org/10.1111/j.1461-0248.2009.01352.x
Manzano Baena P, Casas R (2010) Past, present and future of trashumancia in Spain: nomadism in a developed country. Pastor Res Policy Pract 1(1):72–90.
Manzano P, Burgas D, Cadahía L, Eronen J, Fernández-Llamazares Á, Bencherif S, Holand Ø, Seitsonen O, Byambaa B, Fortelius M, Fernández-Giménez ME, Galvin KA, Cabeza M, Stenseth NC (2021) Toward a holistic understanding of pastoralism. One Earth 4:651–665. https://doi.org/10.1016/j.oneear.2021.04.012
Manzano P, Azcarate FM, Bencherif S, Burgas D, Byambaa B, Cabeza M, Cadahía L, Chatty D, Eronen JT, Galvin K.A, Herrera PM, Holand Ø, Itani M, Niamir-Fuller M, Pauné F, Perrier G, Scoones I, Seitsonen O, Stenseth NC, Varela E, Waters-Bayer A (2023) Grazing research should consider mobility and governance. Sci E-letter 378:6622
Manzano P (2024) El cambio climático y la trashumancia: elementos clave del Plan de Salvaguarda. Ministerio de Cultura de España, Madrid
Manzano P, de Aragão Pereira MA, Windisch W (2025) Vast extension but positive outcomes, reduced but negative: complexity and nuances in evaluating land use by livestock and crops. Anim Front 15:43–54. https://doi.org/10.1093/af/vfae051
Mduma SA, Sinclair ARE, Hilborn R (1999) Food regulates the Serengeti wildebeest: a 40-year record. J Anim Ecol 68:1101–1122. https://doi.org/10.1046/j.1365-2656.1999.00352.x
Parra SA, Ramos-Font ME, Buisson E, Robles AB, Vidaller C, Pavon D, Baldy V, Dominguez P, Godoy-Sepúlveda F, Mazurek H, Peña-Enguix A, Sanosa-Cols P, Corcket E, Genin D (2025) How transhumance and pastoral commons shape plant community structure and composition. Rangel Ecol Manag 98:269–282. https://doi.org/10.1016/j.rama.2024.10.002

Pauné Fabré F (2021) Bases ecològiques per a una gestió pastoral. Doctoral dissertation. Universitat de Lleida, Lleida

Pausas JG, Bond WJ (2020) On the three major recycling pathways in terrestrial ecosystems. Trends Ecol Evol 35:767–775. https://doi.org/10.1016/j.tree.2020.04.004

Perevolotsky A, Seligman NG (1998) Role of grazing in Mediterranean rangeland ecosystems – inversion of a paradigm. Bioscience 48:1007–1017. https://doi.org/10.1007/s10980-023-01676-0

Poore J, Nemecek T (2018) Reducing food's environmental impacts through producers and consumers. Science 360:987–992. https://doi.org/10.1126/science.aaq0216

Riginos C (2015) Climate and the landscape of fear in an African savanna. J Anim Ecol 84:124–133. https://doi.org/10.1111/1365-2656.12262

Serrano-Zulueta R, Pardo G, Pauné F, del Prado A, Manzano P (2023) Herbivory baseline estimates in Spanish protected areas, and environmental implications. Landsc Ecol 38:3713–3729. https://doi.org/10.1007/s10980-023-01783-y

Sinclair AR, Packer C, Mduma SA, Fryxell JM (2009) Serengeti III: human impacts on ecosystem dynamics. University of Chicago Press, Chicago, IL

Velamazán M, Sánchez-Zapata JA, Moral-Herrero R, Jacquemin EG, Sáez-Tovar JA, Barbosa JM (2023) Contrasting effects of wild and domestic ungulates on fine-scale responses of vegetation to climate and herbivory. Landsc Ecol 38:3463–3478

Chapter 10
The Dehesa: A Cultural Landscape Facing Future Challenges

Susanne Schnabel, Manuel Pulido-Fernández and J. Francisco Lavado-Contador

Abstract The dehesa (or montado) is a traditional agrosilvopastoral agroforestry system in the southwestern Iberian Peninsula, characterized by scattered Quercus spp. trees, native livestock breeds, and a diverse herbaceous layer. Recognized for its high natural, cultural, and economic value, the dehesa provides crucial ecosystem services, supports biodiversity, and represents a unique cultural landscape. However, this system faces significant contemporary challenges that threaten its long-term sustainability. These include land degradation (e.g., soil erosion, tree mortality), a critical lack of tree regeneration, and pressures from socio-economic changes such as the abandonment of traditional practices, agricultural intensification, and the impacts of climate change. This chapter synthesizes the dehesa's ecological and socio-economic characteristics, explores its inherent complexities, and the principal drivers of its degradation. It emphasizes the importance of understanding the dehesa as a complex socio-ecological system where management practices are integral to its structure and function. Critical factors are highlighted such as tree regeneration failure, soil health, and the impact of agricultural policies, while some strategies for sustainable management are considered, including facilitating tree regeneration, promoting traditional uses, controlling diseases, and implementing economic incentives for ecosystem services.

Keywords Agroforestry system · Silvopastoral · Grazing · Land degradation · Land use change

S. Schnabel (✉) · M. Pulido-Fernández · J. F. Lavado-Contador
INTERRA Research Institute, Universidad de Extremadura, Cáceres, Spain
e-mail: schnabel@unex.es

M. Pulido-Fernández
e-mail: mapulidof@unex.es

J. F. Lavado-Contador
e-mail: frlavado@unex.es

J. Martínez-Valderrama and J. Olcina Cantos (eds.), *The Labyrinth of Desertification*,
https://doi.org/10.1007/978-3-032-09648-7_10

An Open Woodland with Multiple Uses

The dehesa, or montado in Portuguese, is a traditional agrosilvopastoral system typical of the southwestern Iberian Peninsula, representing a distinct and valuable form of agroforestry. It is primarily based on the extensive rearing of native livestock breeds, mainly for meat production, including sheep, cattle, and pigs, such as the Merino sheep, Retinta cattle, or Iberian pig. These animals graze on pastures year-round and feed on acorns during the "montanera" period (October–February). However, depending on the management of each farm, it is also possible to find goats, fighting bulls, horses, donkeys, chickens, beehives, and game species. Similarly, agriculture can also be present in the form of rotational crops of forage cereals, and there is forest production through cork extraction, pruning for firewood, and mushroom harvesting (Pulido and Picardo 2010).

From a landscape perspective, in its most genuine form, the dehesa is the result of the degradation of Mediterranean woodland through thinning and clearing. This has created a landscape where scattered mature trees of the *Quercus* genus, primarily holm oaks and cork oaks, coexist with a diverse herbaceous layer dominated by annual grasses (Guzmán-Álvarez and Guimarães 2021) (Fig. 10.1). However, some oak groves and even olive or wild olive trees may be subjected to similar uses and management, resulting in dehesa-like landscapes with similar structure and function.

Fig. 10.1 Typical dehesa landscape in spring with mature holm oaks and annual grasses. Photo by Susanne Schnabel

The origin of this degradation remains uncertain, with pollen evidence suggesting woodland thinning in prehistoric times (López Sáez et al. 2007). Nevertheless, there is consensus on the medieval origin of the dehesa (Clemente Ramos 2012) and its subsequent consolidation in later centuries (Llorente Pinto 2003), linked to the Christian reconquest of the territory.

The Present Situation of the Dehesa

Currently, dehesas are mostly large private farms dedicated almost exclusively to extensive livestock farming (Fig. 10.2). Including the Portuguese montados, they cover an area of around 3.5 million hectares, with Extremadura being the most representative region in this regard, with 1.4 million hectares (Fig. 10.3). In Extremadura's dehesas, there is an extensive livestock population of over 3 million sheep, 1.5 million cattle, and 1.5 million Iberian pigs, representing a ratio of 1:6 with respect to the human population (Gaspar et al. 2007). The dehesa, as a unique landscape of the Iberian territory (Mata and Sanz 2003), is more than just a livestock farming system, undoubtedly the most important land use in dehesas. Traditionally, these spaces have had the so-called "*suelo y vuelo*" use, in other words, an example of agrosilvopastoral use adapted to the climatic and edaphic characteristics of some Iberian territories; embodying key principles of agroforestry by ensuring an integral use of the tree (*Quercus*), with acorns used for livestock feed, pruned branches for farmhouse kitchens and stoves, and trunks for construction, when mature trees had

Fig. 10.2 A herd of beef cattle in a dehesa. Photo by Susanne Schnabel

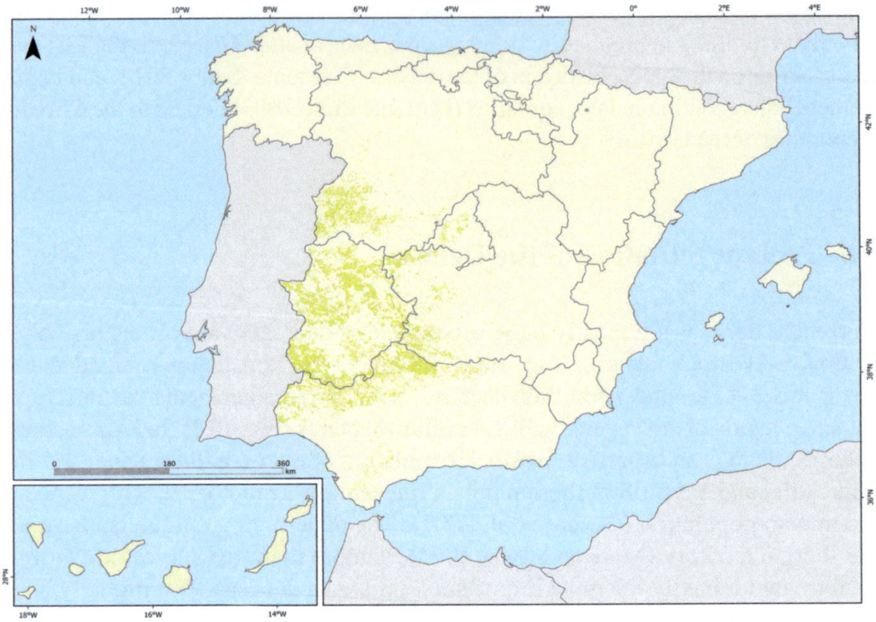

Fig. 10.3 Spatial distribution of dehesas in Spain. Source: Spanish Forest Map (MITECO 2006)

to be cut. Additionally, if annual rainfall conditions allowed, some areas of the farms could be sown for grain (Martin Galindo 1966). The type of pruning applied to the trees was important for establishing the formal structure of the dehesa and the use of wood and acorns (García Fernández 1977).

However, the dehesa should be understood as a complex system of socio-ecological interrelationships, where different agricultural production and livestock management strategies (traditional, conventional, organic, holistic, game management, etc.) converge with various elements of the physical and biological context: soil, water, climate, biota (Plieninger et al. 2021). These elements maintain a fragile relationship, and the deterioration of any one of them can lead to the collapse of the entire system (Moreno et al. 2016). For example, increasing the productivity of a farm at the expense of exceeding the livestock carrying capacity can, in the short term, lead to soil degradation and, in the medium and long term, reduce the production and quality of pastures; as well as worsen the phytosanitary condition of the trees (Pulido et al. 2018).

The report on habitat 6310 "Perennial Quercus spp. dehesas" (Díaz Esteban and Pulido Díaz 2009) provides a comprehensive overview of the characteristics, conservation status, and recommendations for this habitat. It identifies several threats related to the abandonment of traditional practices, overgrazing, land use conversion, and occurrence of diseases and pests. At a broader scale, silvopastoral systems like the dehesa, which are dominant in the Mediterranean region, have experienced a general reduction in their area across Europe between 2009 and 2018. This decline is primarily linked to a decrease in extensive grazing, highlighting the vulnerability of these systems to changes in management practices and agricultural policies (Rubio-Delgado et al. 2023). Therefore, the conservation status of dehesa habitats is mostly classified as unfavourable or inadequate, mainly due to the lack of tree regeneration and threats associated with the abandonment or intensification of agricultural uses. The dynamics of this tree loss, which exacerbates regeneration problems, have been investigated through spatial pattern analysis, identifying a higher incidence in ploughed areas or grasslands compared to shrub-covered areas, and a tendency for clearings to form in intensively used zones (Herguido Sevillano et al. 2017).

Plieninger et al. (2021) identify a series of degradation factors in dehesas, which can be seen as challenges common to many traditional agroforestry systems, summarized as: (1) economic factors, such as declining profitability and pressures to intensify and simplify agricultural production; (2) agricultural policies, such as the CAP, which have incentivized agricultural intensification, increased livestock numbers, and the abandonment of multifunctional management practices (Gonzalo Langa 2011); (3) climate change, which increases the frequency and intensity of forest fires and droughts; and (4) sociocultural changes, such as rural depopulation, aging populations, and reduced availability of labour. In terms of specific pressures on the system, aspects related to the simplification of land uses or changes in forest management and practices are also identified.

Box 10.1: *Dehesas*, a "Desertification Landscape" in Spain's 2008 National Action Programme to Combat Desertification

Agrosilvopastoral systems are one of the five desertification landscapes identified in the National Action Programme to Combat Desertification (MAGRAMA 2008). The increase in livestock pressure, derived from the profitability of the Common Agricultural Policy at the time (these scenarios were formulated in 1998), could lead to erosion processes due to overgrazing. The diagram represents these relationships. Indeed, vegetation cover is essential for erosion control, as it acts as a protective shield against rainfall. In a balanced ecosystem, water leads to increased primary productivity, which livestock convert into protein. If market conditions or subsidy policies are favourable, this profitability attracts new herders and flocks, reducing vegetation cover and triggering erosion processes that reinforce themselves (less vegetation cover -> more erosion -> less soil -> less primary productivity -> less vegetation cover).

Currently, this landscape seems like a deactivated threat, as extensive livestock farming usually corresponds to low livestock pressure systems, which are limited by the climatic characteristics of arid or semi-arid areas, where seasonal and low-productivity pastures predominate, to which only highly resilient native species can adapt. This extensive livestock farming system, deeply rooted in many Spanish regions, if properly managed with appropriate livestock pressure, integrates perfectly into the environment, and not only has no negative effect but also helps maintain ecosystems with a strong tendency towards erosion, as the contribution of organic matter and nutrients to the soil favours the maintenance of vegetation cover. However, there are residual local situations where overgrazing still occurs, especially with sheep and goats. Therefore, although extensive livestock farming in Spain is generally a latent threat, in certain contexts it can reactivate and lead to erosion processes or the loss of palatable species.

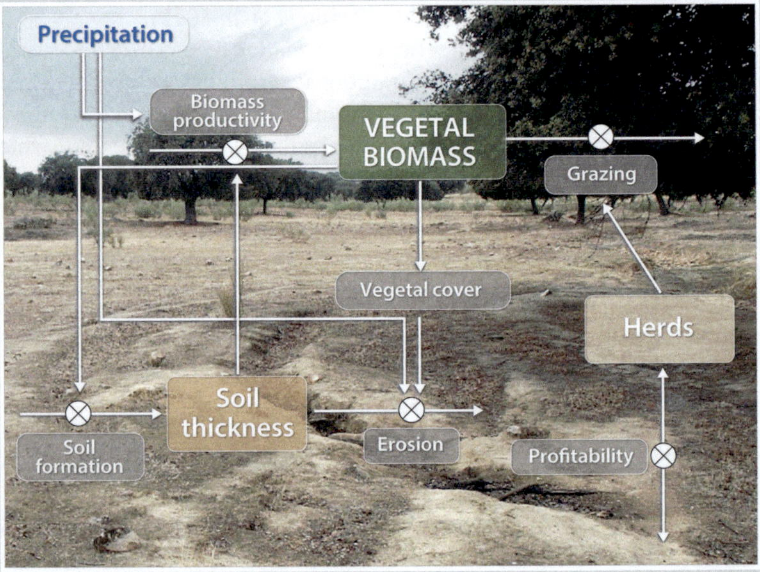

Key interactions within the dehesa socioecosystem, combining ecological and human-managed components

Currently, soil degradation, in the form of erosion, compaction, and loss of fertility (Fig. 10.4), along with increased tree mortality (Fig. 10.5), are the main issues that seem to concern society. However, this negative perception should be relativized, as erosion rates in dehesas are substantially lower than in croplands (Schnabel et al. 2009), holm oaks behave as true fertility islands (Simón et al. 2013), and dehesas show much higher biodiversity values than other agricultural environments with monocropping (Lozano Fondón et al. 2020).

Obviously, if the dehesa is compared to other forest spaces, it can always be interpreted as the result of an anthropogenic degradation process; nevertheless, it undoubtedly represents one of the most appropriate land uses, considering the natural limitations of its geographical context. It provides economic production, animal welfare, protects against the threat of forest fires by reducing shrub encroachment and the flammability of the landscape, offers greater biodiversity than a monospecific forest, and, in addition, suppose a great cultural legacy. In other words, it is a system that can be considered a paradigm of environmental sustainability and even economic

Fig. 10.4 Evidence of soil erosion and degradation of the herbaceous cover in a dehesa with excessive livestock pressure. Photo by Susanne Schnabel

Fig. 10.5 Holm oak affected by "seca" (oak dieback) in its initial phase. Photo by Valentín Maya Blanco

viability, if it is correctly managed and adopting the necessary measures that avoid soil degradation and tree fossilization.

In this sense, the dehesa is considered a biodiversity hotspot, particularly in its herbaceous layer (Guerrero-Casado et al. 2013), stands out for its importance in carbon sequestration (Reyes-Palomo et al. 2022), significantly enhances air and water quality (Calvete Sogo 2016), and provides the local population with exceptional quality food products, as well as recreational areas (Garrido et al. 2017). Additionally, it acts as a protective barrier against natural disasters such as forest fires and problems caused by the overpopulation of game species (Parra-López et al. 2023).

Looking for Sustainability

However, the sustainability of the dehesa requires integrated approaches that consider both ecological and socioeconomic aspects. Future research and management efforts should focus on developing sustainable practices that maintain the multifunctionality of dehesas, while adapting to changing environmental and socioeconomic conditions. To this end, Díaz and Pulido (2009) suggest: (i) Facilitating tree regeneration; (ii) Promoting sustainable management of traditional uses, for example, by encouraging extensive grazing and crop rotation, which help maintain the mosaic structure of the landscape and biodiversity; (iii) Controlling diseases and pests, for example, by implementing comprehensive strategies against "seca," caused by factors such as *Phytophthora cinnamomi* infection, prolonged droughts, and improper pruning; and (iv) Promoting policies that introduce economic incentives and payments for ecosystem services to ensure the viability of traditional practices that maintain the dehesas.

In the short term, to address the most urgent needs that dehesas are facing, aspects such as reducing tillage through direct seeding in areas where cultivation is necessary; intelligent grazing practices, where livestock pressure and new technologies are adapted to the local farm contexts; greater soil protection based on erosion prevention and the construction of control elements, in cases where erosion is inevitable; effective tree regeneration through temporary grazing exclusion from small areas of the farms (e.g., rotations of 5% of the surface every 20 years); and finally, the search for financial compensation mechanisms that favour the economic viability of these farms.

In summary, dehesas are recognized for their high natural, cultural, and economic value, providing important ecosystem services, and promoting biodiversity. Despite the challenges they face, such as intensification, the abandonment of traditional practices, and the effects of climate change, sustainable management can consolidate them as a model of environmental and economic sustainability. To ensure their multifunctionality and resilience, it is essential to adopt integrated strategies that respond to current environmental and socioeconomic dynamics (Plieninger and Wilbrand 2001).

References

Calvete Sogo H (2016) Productivity and quality of dehesa pastures in relation to global change: Increasing tropospheric ozone and nitrogen deposition. Universidad Politécnica de Madrid, Madrid

Clemente Ramos J (2012) Ocupación del espacio y formas de explotación: Dehesas y adehesamiento en Extremadura (c.1250 – c.1450). In: García Fitz F, Jiménez Alcázar JF (eds) La historia peninsular en los espacios de frontera: Las "extremaduras históricas" y la "transierra" (siglos XI-XV). Sociedad Española de Estudios Medievales, Editum, pp 253–277

Díaz Esteban M, Pulido Díaz FJ (2009) 6310 Dehesas perennifolias de Quercus spp. In: V.V.A.A. (ed) Bases ecológicas preliminares para la conservación de los tipos de hábitat de interés comunitario en España. Ministerio de Medio Ambiente y Medio Rural y Marino, p 69

García Fernández J (1977) Formas de explotación. In: Anes Álvarez G, Bernal Rodríguez A, García Fernández J, Giralt Raventós E, Vilar P et al (eds) La economía agraria en la historia de España. Propiedad, explotación, comercialización, rentas, Ediciones Alfaguara, Fundación March, pp 189–210

Garrido P, Elbakidze M, Angelstam P, Plieninger T, Pulido F, Moreno G (2017) Stakeholder perspectives of wood-pasture ecosystem services: a case study from Iberian dehesas. Land Use Policy 60:324–333. https://doi.org/10.1016/j.landusepol.2016.10.022

Gaspar P, Mesías F, Escribano M, Rodriguez de Ledesma A, Pulido F (2007) Economic and management characterization of dehesa farms: implications for their sustainability. Agrofor Syst 71:151–162. https://doi.org/10.1007/s10457-007-9081-6

Gonzalo Langa J (2011) El impacto de la aplicación de la PAC en las producciones ganaderas de la Dehesa (1986-2010). In: Coleto Martínez JM, de Muslera Pardo E, González Blanco R, Pulido García F (eds) La agricultura y la ganadería extremeñas: Informe 2010. Caja de Ahorros de Badajoz, pp 181–196

Guerrero-Casado J, Carpio AJ, Ruiz-Aizpurua L, Tortosa FS (2013) Restocking a keystone species in a biodiversity hotspot: recovering the European rabbit on a landscape scale. J Nat Conserv 21(6):444–448. https://doi.org/10.1016/j.jnc.2013.07.006

Guzmán-Álvarez JR, Guimarães MH (2021) The spatial, temporal and social construction of the concept of silvopastoral systems: the case of montado and dehesa. In: Pinto-Correia T, Guimarães MH, Moreno G, Acosta-Naranjo R (eds) Governance for Mediterranean Silvopastoral Systems. Routledge, London, pp 127–141

Herguido Sevillano E, Lavado Contador JF, Pulido M, Schnabel S (2017) Spatial patterns of lost and remaining trees in the Iberian wooded rangelands. Appl Geogr 87:170–183. https://doi.org/10.1016/j.apgeog.2017.08.011

Llorente Pinto JM (2003) El problema de la sostenibilidad de las dehesas a la luz de la evolución histórica de los terrenos adehesados. Cuad Soc Esp Cienc Forestales 16:135–140

López Sáez JA, López García P, López Merino L, Cerrillo Cuenca E, González Cordero A, Prada Gallardo A (2007) Origen prehistórico de la dehesa en Extremadura: Una perspectiva paleoambiental. Rev Estudios Extremeños 63(1):493–510

Lozano Fondón C, Barrena-González J, Pulido Fernández M, Remelli S, Lozano Parra J, Menta C (2020) Effects of livestock pressure and vegetation cover on the spatial and temporal structure of soil microarthropod communities in Iberian rangelands. Forests 11(6):628. https://doi.org/10.3390/f11060628

Martín Galindo JL (1966) La dehesa extremeña como tipo de explotación agraria. Estudios Geográficos 27(103):157–247

Mata Olmo R, Sanz Herranz C (2003) Atlas de los paisajes españoles. Ministerio de Medio Ambiente, Madrid, p 683

MAGRAMA (2008) Programa de Acción Nacional contra la Desertificación.Ministerio de Agricultura y Medio Ambiente, Madrid, Spain

MITECO (Ministerio para la Transición Ecológica y el Reto Demográfico) (2006) Mapa Forestal de España 1:50.000 (MFE50). Dirección General de Biodiversidad, Subdirección General de

Política Forestal, Madrid. Retrieved from https://www.miteco.gob.es/en/biodiversidad/temas/inventarios-nacionales/inventario-forestal-nacional/mfe50.aspx [Accessed: June 20, 2025]

Moreno G, González-Bornay G, Pulido F et al (2016) Exploring the causes of high biodiversity of Iberian dehesas: the importance of wood pastures and marginal habitats. Agrofor Syst 90:87–105. https://doi.org/10.1007/s10457-015-9817-7

Parra-López C, Sayadi S, Garcia-Garcia G, Abdallah SB, Carmona-Torres C (2023) Prioritising conservation actions towards the sustainability of the dehesa by integrating the demands of society. Agric Syst 206:103613. https://doi.org/10.1016/j.agsy.2023.103613

Plieninger T, Flinzberger L, Hetman M et al (2021) Dehesas as high nature value farming systems: a social-ecological synthesis of drivers, pressures, state, impacts, and responses. Ecol Soc 26(3):23. https://doi.org/10.5751/ES-12647-260323

Plieninger T, Wilbrand C (2001) Land use, biodiversity conservation, and rural development in the dehesas of Cuatro Lugares, Spain. Agrofor Syst 51(1):23

Pulido F, Picardo Á (2010) Libro verde de la dehesa. Junta de Castilla y León, SECF, SEEP, AEET, SEO

Pulido M, Schnabel S, Lavado Contador JF, Lozano-Parra J, González F (2018) The impact of heavy grazing on soil quality and pasture production in rangelands of SW Spain. Land Degrad Dev 29(2):219–230. https://doi.org/10.1002/ldr.2501

Reyes-Palomo C, Aguilera E, Llorente M, Díaz-Gaona C, Moreno G, Rodríguez-Estévez V (2022) Carbon sequestration offsets a large share of GHG emissions in dehesa cattle production. J Clean Prod 358:131918. https://doi.org/10.1016/j.jclepro.2022.131918

Rubio-Delgado J, Schnabel S, Burgess PJ, Burbi S (2023) Reduced grazing and changes in the area of agroforestry in Europe. Frontiers in Environmental Science 11:1258697. https://doi.org/10.3389/fenvs.2023.1258697

Schnabel S, Gómez Gutiérrez Á, Lavado Contador J (2009) Grazing and soil erosion in dehesas of SW Spain. In: Romero Díaz, MA, Belmonte Serrano F (eds) Advances in studies on desertification. Editum, Murcia, pp 725–728

Simón N, Montes F, Díaz-Pinés E, Benavides R, Roig S, Rubio A (2013) Spatial distribution of the soil organic carbon pool in a Holm oak dehesa in Spain. Plant Soil 366(1):537–549. https://doi.org/10.1007/s11104-012-1443-9

Chapter 11
Desertification Due to Agricultural Land Abandonment

Estela Nadal-Romero and Teodoro Lasanta

Abstract Land abandonment is a widespread phenomenon in Spain, particularly in Mediterranean mountain areas and semi-arid regions, where it constitutes a key driver of land degradation and desertification. This chapter analyses the environmental consequences of farmland abandonment, with a specific focus on soil erosion and conservation processes. While land abandonment can initiate ecological succession and lead to the recovery of soil quality, this outcome is not guaranteed. In many cases, degradation processes persist or intensify, especially where cultivation had previously exhausted soil fertility, where traditional soil and water conservation practices, such as terracing, have been abandoned, or where recurrent wildfires inhibit vegetation regrowth. In semi-arid environments, low rainfall and high soil erodibility further constrain vegetation recovery, facilitating crust formation, runoff concentration, and gully development. Drawing on evidence from various Spanish regions, the study underscores the importance of site-specific management strategies aimed at mitigating the adverse impacts of abandonment and enhancing ecosystem services. Without appropriate interventions, land abandonment may exacerbate erosion, reduce biodiversity, and increase the risk of irreversible degradation, compromising long-term sustainability in vulnerable landscapes.

Keywords Land abandonment · Mediterranean mountain areas · Land degradation · Semi-arid regions · Soil erosion · Terraces

E. Nadal-Romero (✉) · T. Lasanta
Instituto Pirenaico de Ecología (IPE-CSIC), CSIC, Zaragoza, Spain
e-mail: estelanr@ipe.csic.es

T. Lasanta
fm@ipe.csic.es

J. Martínez-Valderrama and J. Olcina Cantos (eds.), *The Labyrinth of Desertification*,
https://doi.org/10.1007/978-3-032-09648-7_11

An Unexpected Landscape of Desertification

At the end of 1996, the United Nations Convention to Combat Desertification (UNCCD 1994) came into force to address land degradation. Spain ratified this Convention, which required the country to develop and implement a National Action Programme to Combat Desertification (PAND). This programme was launched in August 2008 and identifies the main factors and processes contributing to desertification in the northern Mediterranean region, including 'the crisis conditions in traditional agriculture, leading to land abandonment, soil deterioration, and the degradation of water conservation structures'. According to the assessment presented in the PAND (2008), 17.8% of Spain's territory is at high or very high risk of desertification, coinciding with areas prone to severe erosion, forest fires, and/or soil salinisation (García-Ruiz and López Bermúdez 2009).

The term desertification is complex and is used from different perspectives. However, interpretations converge on the idea that it involves a virtually irreversible decline, on a human timescale, in the productivity levels of terrestrial ecosystems. Desertification represents the final stage of environmental degradation, characterised by the loss of natural vegetation cover and the intensification of erosive processes. These lead to soil loss, a decline in fertility and productive capacity, a reduction in biodiversity, and the disruption of ecological balances.

Desertification is the result of a combination of geographical, climatic, and socio-economic factors, with human activities playing a particularly significant role. These activities have been identified as the primary drivers of land degradation and desertification in Spain (García-Ruiz and López García 1997). One of the most critical anthropogenic factors is the abandonment of agricultural land. In addition to land degradation, this process leads to the loss of ecosystem goods and services, jeopardising the sustainability of certain societies (García Llamas et al. 2019).

The following pages define what is meant by land abandonment and examine its underlying causes. This is followed by an analysis of how land abandonment contributes to desertification in Spain, with a particular focus on Mediterranean mountain regions and semi-arid environments.

Land Abandonment and Its Causes in Spain

There are different definitions of farmland abandonment, depending on the affected area, the context of the study, and its environmental and socio-economic implications. However, a clear and simple definition is provided by Romero Díaz (2016): 'the phenomenon of the permanent cessation of agricultural activity in a given territory'.

The abandonment of agricultural land is a global phenomenon that particularly affects developed countries, with Spain being one of the most affected. According to FEGA (2019), there are 2.32 million hectares of abandoned land in Spain, representing 20% of the cultivated area and 4.5% of the total agricultural land. Mediterranean mountain regions and semi-arid drylands are the most affected areas (García-Ruiz and Lana-Renault 2011).

Several causes have led to land abandonment, according to Lasanta et al. (2019): (i) rural depopulation; (ii) technological innovations and the mechanisation of agricultural tasks, which have displaced labour from the countryside, and made cultivation on certain slopes impossible; (iii) the low productivity of certain fields; (iv) the high competitiveness of markets; (v) the impact of the Common Agricultural Policy (CAP), which encourage land abandonment between 1989 and 2008; (vi) climatic conditions in semi-arid environments, due to the high erosivity and irregular rainfall; (vii) the availability of alternative employment outside agriculture; (viii) factors related to management: overexploitation and depletion of the land, salinisation due to irrigation, soil contamination, and abandonment of traditional soil conservation practices and structures (such as terraces). In summary, the causes are multiple and include socio-economic, environmental, and management factors.

Soil Erosion/Conservation in the Mediterranean Mountains

Most of the slopes in the Spanish mountains were abandoned during the twentieth century as a result of depopulation and the mechanisation of agricultural work. Today, in the best cases, only those plots of land that are suitable for machinery are cultivated: they have access roads, have a slope of less than 10%, and are large enough for tractors to manoeuvre. The result has been a massive abandonment of fields; by the year 2000, the formerly cultivated slopes were almost completely abandoned, or nearly so, in the north-western Iberian System, the Pyrenees, Galicia, the Cantabrian Coast, the Central System, and a large part of the Betic Mountains. Since the beginning of this century, land abandonment has continued, although at a slower rate (Lasanta et al. 2021).

Thus, land abandonment is widespread in the Spanish mountains. However, it is in the Mediterranean and sub-Mediterranean mountains that the risks of desertification are greatest due to the higher levels of past cultivation of marginal lands, the slow rate of vegetation succession following abandonment and the greater erosivity of rainfall. In the Atlantic mountain ranges, vegetation quickly covers abandoned fields, reducing runoff and soil loss.

When a field is abandoned, processes of plant succession begin, which generally involve a reduction in runoff rates and sediment transport compared to the cultivation phase. Soil properties (such as bulk density, fertility, aggregate stability, organic matter content, etc.) tend to improve a few years after abandonment, with increased water infiltration. In short, processes of edaphogenesis occur, leading to an improvement in soil quality. However, this is not always the case. There are three situations in which soil quality does not improve, but rather degradation processes occur that can lead to desertification:

1. When the soil of the abandoned field (or in many cases of the entire slope) has been depleted during the cultivation phase, leaving it weak and of low fertility. This was the case, for example, with the practice of slash-and-burn agriculture

(shifting agriculture) in the Spanish Mediterranean mountains until the early decades of the twentieth century. In these cases, very steep slopes were used to grow cereals, resulting in high rates of erosion that left the soil impoverished, infertile, and stony. In this situation, plant colonisation is very slow (Fig. 11.1), and soil losses remain very high for decades (Lasanta et al. 2017).

2. The abandonment of water and soil conservation practices, such as in terraced slopes, is another factor. In terraced fields, the dominant erosion process is the collapse of the terrace walls (Fig. 11.2), which creates scars that degrade the landscape and promote the formation of gullies, leading to the export of large amounts of sediment (Arnáez et al. 2015). The presence of soil, sometimes several metres thick, facilitates the collapse of the terrace walls, resulting in the fall of part of the retaining wall and the transport of soil downhill. Terraces were originally built to promote water infiltration and prevent erosion. It is undoubtedly an excellent technique for reducing erosion on cultivated slopes by reducing the slope gradient, hydrological connectivity, and runoff. However, regular maintenance is essential: repairing collapsed walls after a storm, maintaining drainage systems both inside and outside the fields to channel runoff water into gullies and streams. It should not be forgotten that the more a slope is modified (and terracing is undoubtedly such a modification), the more energy is required to maintain it.

Fig. 11.1 Slash-and-burn slope in Aísa (Central Pyrenees). Almost a century after abandonment, the vegetation cover is very poor, with signs of erosion processes. Photo by Teodoro Lasanta

Fig. 11.2 Collapses in stone walls on abandoned terraces in San Martín del Jubera (La Rioja). Photo by Teodoro Lasanta

3. The occurrence of repeated fires. Plant colonisation leads to an increase in biomass. In the Mediterranean climate, natural fires occur with some frequency in addition to those caused by human activity. In areas subject to recurrent fires, plant colonisation is severely hampered, leading to progressive soil degradation. In the Pyrenees, for example, it has been observed that many south-facing slopes (Fig. 11.3) have been cultivated and then periodically burned to regenerate pastures. Several decades later, these slopes have highly degraded soils and a very poor vegetation cover, consisting of stunted shrubs and isolated patches of bare soil.

Soil Degradation in Semi-Arid Environments

The South-eastern Iberian region (the southern half of the province of Alicante, the eastern sector of Albacete, the provinces of Murcia and Almería, and part of Granada) and the central area of the Ebro depression (parts of the provinces of Zaragoza, Huesca, and Navarra) receive less than 350 mm of average annual rainfall

Fig. 11.3 Slope of abandoned fields in the Borau valley (Central Pyrenees). The boundaries of the former fields and the slow encroachment of shrub vegetation are visible. Photo by Teodoro Lasanta

(with very irregular rainfall both annually and interannually) and have average temperatures exceeding 14°C. Droughts are frequent and prolonged, and the water balance is very negative throughout the year. This area has a semi-arid climate, with poorly structured and fragile soils (sodic and/or saline) that show clear signs of desertification.

In semi-arid environments, plant succession after abandonment is very slow (Fig. 11.4), leading to a progressive loss of nutrients and organic matter. In these areas, soil degradation is closely linked to water availability and its effect on the spatial distribution of vegetation cover. Where some moisture accumulates, plants grow and improve soil quality, while in bare areas are subject to erosion processes such as sheet erosion, diffuse erosion, rills, and gullying. Another important factor is the formation of crusts after abandonment. During the cultivation phase, ploughing removes crusts and facilitates water infiltration. Immediately after abandonment, the soil becomes compacted, loses porosity, and surface crusts form, reducing infiltration and increasing runoff and erosion (Romero Díaz 2016).

In addition to soil moisture, lithology has a strong influence on plant succession. In the Region of Murcia, Romero Díaz et al. (2007) found that abandoned fields on calcareous substrates are rapidly colonised by herbaceous plants, which soon give

Fig. 11.4 Abandoned fields in Mediana (Zaragoza, Spain). The plots were terraced to promote water infiltration. After abandonment, the difficulty for herbaceous plants and shrubs to colonise the fields is evident. Photo by Teodoro Lasanta

way to a shrub cover, increasing organic matter and promoting soil conservation and landscape diversity. On the other hand, abandonment on marl substrates encourages the formation of surface crusts, reducing water infiltration. At the same time, the soil cracks, facilitating the channelling of both surface and subsurface runoff, creating a dense network of gullies, often guided by the collapse of pipes (Romero Díaz et al. 2007), leading to desertification processes (Fig. 11.5).

It is clear from the above that on abandoned land, environmental conditions and management practices can either lead to soil degradation or improve soil quality. Thus, abandonment processes that do not implement appropriate conservation measures tailored to the specific situation contribute to land degradation and desertification. Therefore, Lasanta et al. (2019) suggest that it seems advisable to adopt management practices adapted to the characteristics of the environment to minimise certain negative impacts (such as the fire risk, soil degradation, and landscape homogenisation) and promote the provision of ecosystem goods and services, such as soil conservation, production of high quality water, carbon sequestration, and enhancement of biodiversity. It should not be forgotten that some abandoned fields have soils of sufficient quality to support forestry and agricultural uses. Their productive and environmental potential can drive the reintegration of abandoned land into the socio-economic system.

Fig. 11.5 Pipes in Valpalmas (Zaragoza, Spain). Piping processes can be very prevalent in valley bottoms of semi-arid environments. Photo by Estela Nadal-Romero

Acknowledgements This research was supported by the MOUNTWATER (TED2021-131982BeI00 and MCIN/AEI/doi: 10.13039/ 501100011033), funded by the Spanish Ministry of Science and Innovation and NextGeneration EU/PRTR. The 'Geoenvironmental Processes and Global Change' (E02_23R) research group is financed by the Aragón Government and the European Social Fund (ESF-FSE).

References

Arnáez J, Lana-Renault N, Lasanta T, Ruiz-Flaño P, Castroviejo J (2015) Effects of farming terraces on hydrological and geomorphological processes. A review. Catena 128:122–134. https://doi.org/10.1016/j.catena.2015.01.021

García Llamas P, Geijzendorffer IR, García-Nieto AP, Calvo L, Suárez-Seoane S, Cramer W (2019) Impact of land cover change on ecosystems services supply in mountain systems: a case study in the Cantabrian Mountains (NW of Spain). Reg Environ Change 19:529–542. https://doi.org/10.1007/s10113-018-1419-2

García-Ruiz JM, Lana-Renault N (2011) Hydrological and erosive consequences of farmland abandonment in Europe, with special reference to the Mediterranean region – a review. Agric Ecosyst Environ 140:317–338. https://doi.org/10.1016/j.agee.2011.01.003

García-Ruiz JM, López Bermúdez F (2009) La erosión del suelo en España. Sociedad Española de Geomorfología, Zaragoza, 441 pp

García-Ruiz JM, López García P (1997) Acción humana y desertificación en ambientes mediterráneos. Consejo Superior de Investigaciones Científicas, Instituto Pirenaico de Ecología, Zaragoza, 338 pp
FEGA (2019) Fondo Español de Garantía Agraria. https://www.agronegocios.es/espana-cuenta-con-mas-de-232-mha-de-superficie-agricola-abandonada-y-sin-aprovechamiento/
INCD (1994) International Convention to Combat Desertification. United Nations Environment Programme (UNEP). Interim Secretariat for the Convention to Combat Desertification (CCD). Châtelaine, Geneve, 71 pp
Lasanta T, Arnáez J, Nadal-Romero E (2019) Soil degradation, restoration and management in abandoned and afforestation lands. Adv Chem Pollut Environ Manag Protection 4:71–117. https://doi.org/10.1016/bs.apmp.2019.07.002
Lasanta T, Nadal-Romero E, Errea MP (2017) The footprint of marginal agriculture in the Mediterranean mountain landscape: an analysis of the Central Spanish Pyrenees. Sci Total Environ 599–600:1823–1836. https://doi.org/10.1016/j.scitotenv.2017.05.092
Lasanta T, Nadal-Romero E, Khorchani M, Romero-Díaz MA (2021) Una revisión sobre las tierras abandonadas en España: De los paisajes locales a las estrategias globales de gestión. Cuadernos Investig Geogr 47(2):477–521. https://doi.org/10.18172/cig.4755
PAND (2008) Programa de Acción Nacional contra la Desertificación. Dirección General para la Biodiversidad. Ministerio de Medio Ambiente, Madrid, 131 pp
Romero Díaz MA (2016) Abandono de cultivos en la Región de Murcia. Consecuencias Ecogeomorfológicas. Servicio de Publicaciones de la Universidad de Murcia, Murcia, 263 pp
Romero Díaz MA, Marín-Sanleandro P, Sánchez-Soriano A, Belmonte Serrato F, Faulkner H (2007) The causes of piping in a set of abandoned agricultural terraces in Southeast Spain. Catena 69:282–293. https://doi.org/10.1016/j.catena.2006.07.008
UNCCD (1994) Convención de las Naciones Unidas de Lucha contra la Desertificación. París

Chapter 12
Tourism and Water Stress: A Worrying Convergence in Time and Place

Fernando Gallardo Olmedo

Abstract The Spanish tourism sector is a driving force of the Spanish economy. In 2024, it was the country with the second largest number of foreign tourists in the world (93.8 million), surpassed only by France. Tourism in Spain is, therefore, a productive sector that generates employment and accounted for 13.1% of GDP in 2024. Like other productive sectors, tourism consumes water resources. The analysis carried out in this respect shows that water consumption per person in tourist dwellings is similar to residential consumption, while consumption per person in hotel establishments is approximately four times residential consumption. On the other hand, it is clear that the concentrated tourism model, which is the one followed by Benidorm, generates a lower unitary water consumption than the extended model. Another highly relevant issue is that the Spanish tourist destinations that receive the most visitors are located in areas of very high water stress and, moreover, are concentrated in the summer months, which are the months in which this water stress is most acute. This means that water consumption in the tourism sector must be adjusted in situations where consumption restrictions are necessary.

Keywords Efficient water management · Natural capital · Sustainability · Territorial model · Tourism · Water consumption · Water stress

F. Gallardo Olmedo (✉)
Departamento de Financiación e Investigación Comercial, Universidad Autónoma de Madrid, Madrid, Spain
e-mail: fernando.gallardo@uam.es

J. Martínez-Valderrama and J. Olcina Cantos (eds.), *The Labyrinth of Desertification*,
https://doi.org/10.1007/978-3-032-09648-7_12

Tourism: A Powerful Economic Engine in Spain

The Spanish tourism sector is a driving force of the Spanish economy. It is estimated to have accounted for 13.1% of Spanish GDP by 2024, which is 0.5 percentage points higher than the previous all-time high of 12.60% recorded in 2019 (Statista 2025). Figure 12.1A shows the evolution of this variable since 2016. It can also be seen that the negative impact of the COVID-19 pandemic in 2020 has now been overcome.

Regarding the number of foreign tourists visiting Spain, 2024 also marked a record year, with 93.8 million visitors (INE). Only France surpasses Spain, with a number of 99.8 million visitors that year (EFE 2025). Figure 12.1B shows the historical evolution of this variable, which shows that the break in the series caused by the pandemic has been overcome.

Spain's tourism industry also relies on trips from residents in Spain. In this regard, data are available on the number of trips by Spanish residents within the country itself. As can be seen in Fig. 12.1C, the figure for 2024 is above pre-pandemic levels. Thus, the dynamism of the Spanish tourism sector comes from both residents and non-residents.

The strength of tourism in Spain is due to three factors. First, its natural and cultural capital make Spain highly attractive for tourism. Second, Spain has infrastructure appropriate for the development of tourism, such as advanced transportation,

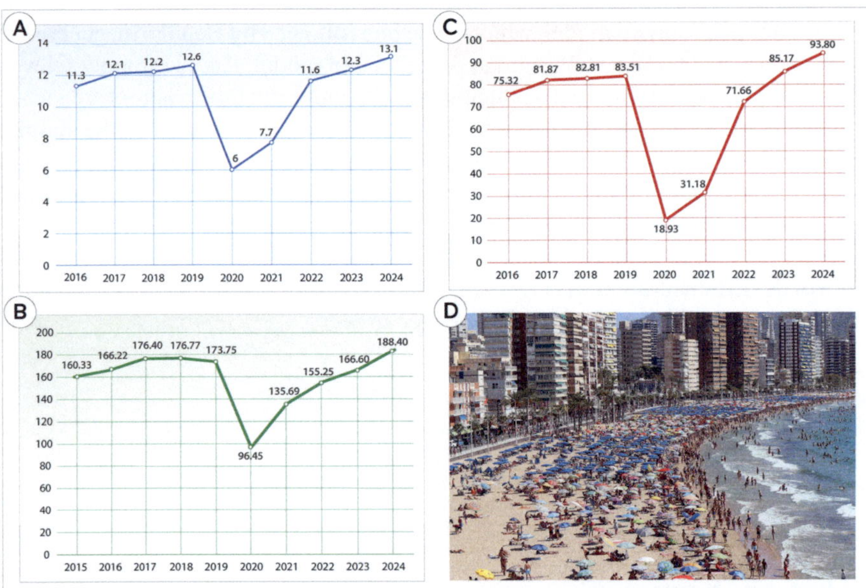

Fig. 12.1 **A** Percentage of GDP contributed by the tourism sector in Spain (Statista 2025). **B** Evolution of the number of foreign tourists who have visited Spain (millions) (own elaboration with data taken from INE 2025a). **C** Evolution of the number of trips made by residents in Spain within the country itself (millions). Data taken from INE (2025b). **D** Crowded beach in Spain during the summer period

healthcare, accommodation, and telecommunications infrastructure of all kinds. Third, the level of security, both in terms of geopolitical risk compared to competing destinations and the safety of Spanish populations, is high. However, there are challenges facing the Spanish tourism sector. In this regard, the following aspects must be considered: the impact of climate change, greenhouse gas emissions and water consumption derived from tourism activities, and the social unrest or *tourismophobia* experienced in some populations (Gallardo 2025).

The Water Footprint of Tourism

As mentioned above, water consumption associated with tourism activities in Spain deserves to be analyzed. The first comment to be made in this regard is that the available data are limited and do not allow for a completely accurate quantification of water consumption for different tourism activities. Taking these limitations into account, we proceed to analyze the available information.

The starting point is the analysis of the total water consumption in Spain. Figure 12.2A shows the total water consumption and its breakdown into four categories (supply, agriculture, industry, and others) for the 2020–2021 hydrological year, which is the latest period available, and for the two preceding years. The largest consumer of water is the agricultural sector, accounting for 77.6% of the total in the 2020–2021 hydrological year, although both the percentage and the amount consumed have been declining. Regarding the supply category, which accounts for water consumption by the tourism sector, the volume of water used is 17% of the total, and its trend in recent years has shown considerable stability. Total water used for consumptive uses comes from two main sources: surface water, with 71.6%; and groundwater, with 23.1%. The remainder comes from reuse, desalination and transfers (Fig. 12.2B).

Regarding the portion of water consumption attributable to tourism activities, INE publishes data for a section called "tourism and recreational uses," which includes hotels, tourist accommodation, campsites, water parks, and golf courses. The evolution of consumption under this heading is shown in Fig. 12.2C. As can be seen, there is no upward trend between 2014 and 2022. The water consumption within this category that generates the most controversy is that related to golf courses. Accurate data on golf course water consumption are not available. A study by López (2024) estimates water consumption and its origin based on surveys conducted at Spanish golf courses. The study concludes that on small courses (with fewer than 18 holes), annual water consumption ranges from 100,000 to 200,000 m^3, with 72% of the water coming from recycled water. On larger courses, with 18 holes, average annual consumption per course ranges from 100,000 to 300,000 m^3, and 57.3% of the water comes from recycled water. It might be assumed that those courses that did not respond to the survey use a higher proportion of non-recycled water.

On the other hand, the wide margins of water consumption are determined by climatic factors related to the geographic location of the golf course. The region with the most golf courses is Andalusia, with 90. Catalonia follows with 44. Both are highly

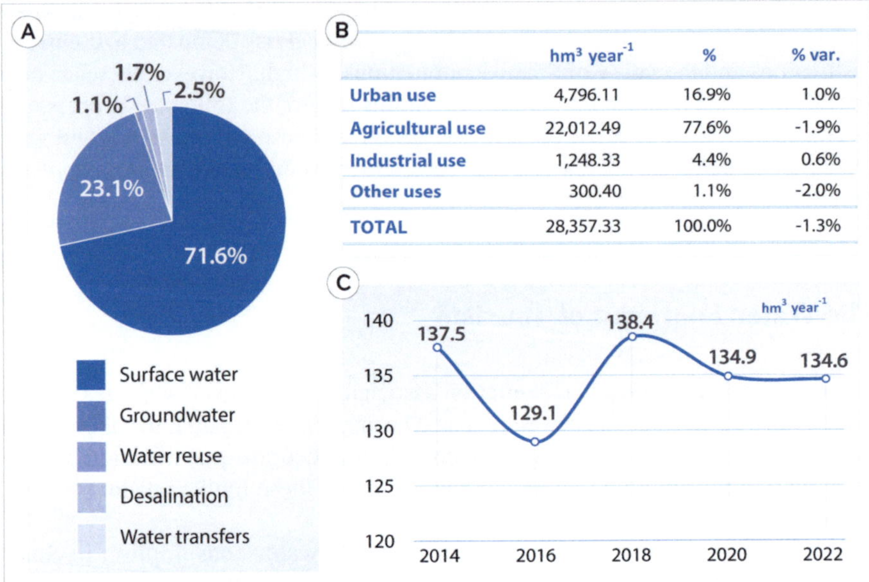

	hm³ year⁻¹	%	% var.
Urban use	4,796.11	16.9%	1.0%
Agricultural use	22,012.49	77.6%	-1.9%
Industrial use	1,248.33	4.4%	0.6%
Other uses	300.40	1.1%	-2.0%
TOTAL	28,357.33	100.0%	-1.3%

Fig. 12.2 **A** Water used by sector. **B** Water used for consumptive uses by origin. Source: MITECO (2022). **C** Evolution of water supplied for tourism and recreational uses (hm^3 year^{-1}). Source: INE (2025c)

touristic regions and do not have the most water resources. However, a complete analysis of this issue should take into account, in addition to the ecological impact and water consumption, the economic contribution generated by the activity in the area where the golf course is located (Gallardo 2025).

However, analyzing the golf course problem requires a broader territorial approach, as is always the case when we talk about desertification. In this regard, the most relevant issue is the urban development model associated with golf. That is, beyond the water used to irrigate the course, the greatest impact is the urbanization associated with the appeal of a golf course. In this sense, the water impact is much greater. Furthermore, the developments linked to this tourism model occupy coastal areas of high ecological value. In the purely economic comparison, tourism obviously wins, but we must recognize that if only the economic criterion were considered, a situation would arise in which the entire coast would be urbanized. This would mean that the natural capital in the form of open spaces, coastal lagoons, birds that pass through the area, and vegetation in the areas would be very negatively affected. It would then be realized that it turned out to have an intangible environmental and economic value that was the true attraction of that area. This means that places in Spain with a unique natural heritage, such as the Cabo de Gata Natural Park in Almería, must be preserved to balance environmental conservation with sustainable economic development (Fig. 12.3).

Fig. 12.3 The Cabo de Gata Natural Park has successfully regulated the tourism that comes to the area. The limitations imposed by the park allow for the preservation of these coastal ecosystems, which act as a natural attraction. If these restrictions were lifted, the coast would be completely urbanized, losing its natural value and likely increasing its income in the short term. Photo by Fernando Gallardo

Droughts and Tourism in Regions with High Water Stress

The statistics corresponding to water used under the heading of "tourist and recreational uses" do not include consumption in the so-called Tourist Dwellings (TD). The number of these dwellings continues to grow in Spain (Fig. 12.4A), except in the provinces of Barcelona and the Balearic Islands, and it is interesting to know their water consumption. Castro et al. (2024) has made an estimate of water consumption in the TDs according to the analysis of 11,000 water bills of 1800 TDs in 2023, and concluded that the average daily consumption per TD was 443 l per day. Taking into account that the average occupancy of the TDs is around 4 people, an average daily consumption of 111 l per person is obtained. According to INE (2024), the average daily consumption per inhabitant in 2022 was 128 l per person, with slight variations by regions (Fig. 12.4B). Therefore, water consumption in the TDs is somewhat lower than the average for Spanish households. However, there is no data available on consumption in TDs by provinces or regions to make a comparison with the average household consumption by province. In any case, it can be stated that there is no comparative waste of water in the TDs, which does not prevent them from generating other types of external effects in the areas where they are located. The aforementioned study also makes a comparison with the water consumed per person in the establishments where they are located and concludes that this is approximately four times that consumed per person in a TD.

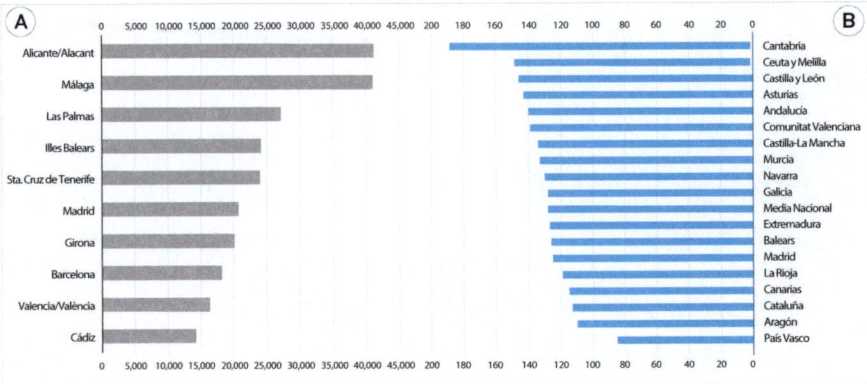

Fig. 12.4 **A** Provinces with the highest number of tourist dwellings (TD) in Spain in 2024 and percentage change compared to 2020. Data Source: INE (2025d). **B** Daily water consumption per capita by regions (2022). Data Source INE (2025e)

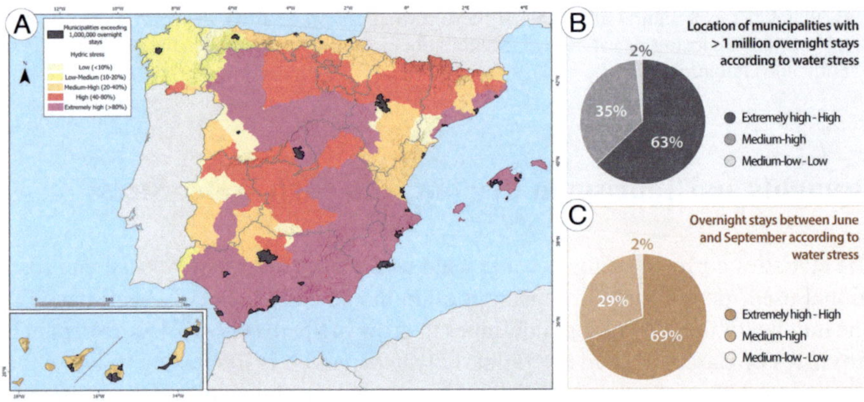

Fig. 12.5 **A** Water stress map of Spain and municipalities with more than one million hotel overnight stays. Source: Kuzma et al. (2023) and INE (2025f). **B** Percentage of populations with more than one million overnight stays per year according to water stress category (INE 2025f). **C** Percentage of overnight stays in the summer period (June to September) according to water stress category

An interesting aspect related to the topic analyzed is to observe where the main Spanish tourist spots are located in relation to water stress, which is defined by Kuzma et al. (2023) as the ratio between the total freshwater used and the total renewable freshwater resources over a period of time. Figure 12.5A shows the water stress map of Spain in 2024 and those municipalities that exceed one million overnight stays per year. As we can see, Spain is subject to high water stress, with extremely high-water stress occurring in 42.8% of the territory, high in 24.9%, medium-high in 19.7%, and low-medium or low in only 7.4%.

As previously mentioned, per capita water consumption in hotels far exceeds that of residents and tourists using the TDs. However, tourists staying in TDs contribute additional water consumption. Therefore, it is interesting to observe whether the municipalities with the most hotel overnight stays and the most TDs are located in areas of water stress. Regarding overnight stays in hotels, it is noteworthy that 24 of the 43 municipalities with the highest number of overnight stays are located in areas with high or extremely high-water stress, many of them in the Mediterranean arc. Furthermore, 12 of these 24 towns concentrate more than 50% of overnight stays in the months of June to September, when stress is greatest. Furthermore, 68% of overnight stays occur in the areas of greatest water stress during the driest time of year, the period between June and September (Fig. 12.5C). Regarding the TD, 8 of the 10 provinces with more than 11,000 TD (Alicante, Málaga, the Balearic Islands, Madrid, Girona, Barcelona, Cádiz, and Tarragona) are located in areas of high or extremely high-water stress.

Territorial Model of Tourism and Water Consumption

After analyzing the available global data on water consumption in tourism activities, it is worth considering a series of considerations related to the construction model of tourist accommodation in different tourist destinations. Three essential factors shape these models: types of water consumption needs, degree of verticality, and distance between buildings.

Regarding the first factor, three consumption needs can be distinguished: water to cover basic needs (food, appliances, and hygiene), water for the swimming pool, and water for garden maintenance. All buildings require water for the first of these, and some for one or both of the remaining components. The degree of verticality refers to the height of the building. If we combine these two factors with the distance between buildings, we find two basic construction models, among which some intermediate ones could be defined. At one extreme is the concentrated model, characterized by many high-rise buildings with multiple accommodations, with shared swimming pools and gardens, and with little distance between them. At the other extreme is the extensive model, with individual buildings equipped with their own swimming pool and garden, and with separation between them. Unit water consumption in the concentrated model is, logically, much lower than that recorded in the extensive model.

In Spain, a consolidated concentrated model stands out, where, in addition, the municipal administration and the hotels in the area have been involved in efficient water use. This is the Alicante town of Benidorm, which was the fourth Spanish city with the highest number of overnight stays in 2024, exceeding the figure of 11.5 million. As Olcina et al. (2016) point out, Benidorm is a tourist destination with a high number of high-rise buildings in close proximity to each other. Furthermore, the hotels in the area have been highly involved in the search for efficient solutions to save water consumption, being pioneers in implementing measures such as efficient faucets, the use of high-efficiency appliances, the installation of efficient pumps, the incorporation of efficiency improvement measures in garden irrigation, and the continuous maintenance of swimming pools, with the consequent water savings. In addition, they have invested in providing access to water during times of scarcity by creating reservoirs through the construction of wells, cisterns, and underground water tanks in some hotels.

Olcina et al. (2016) also highlight in their research the importance of the involvement of the municipal administration in the maintenance of water distribution networks with the aim of minimizing losses due to leaks. The success in terms of water management in Benidorm is evident in three comparisons: (i) in 2022 the average water consumption per inhabitant in Benidorm will not exceed 120 l, which is a figure that is below the national average and the average of the Valencian Community (Hosteltur 2023), (ii) Benidorm occupies the first position among European cities in the Infrastructure Leakage Index prepared by the International Water Association and the third position in the world, only behind Tokyo and Singapore (Hosteltur 2023); and (iii) total water consumption in Benidorm in 2024, with a registered population of around 74,500 inhabitants, was 11 hm^3 (Alicante Plaza 2025) compared to 10 hm^3 in 1975 for a registered population of 15,000 people (Olcina et al. 2016).

The extensive model, as opposed to the concentrated one, has proliferated in quite a few areas of the Spanish Mediterranean (the case of the Costa Brava is a good example, as explained by Gabarda et al. 2015) and especially in the surroundings of golf courses. As mentioned above, golf courses largely use recycled water for irrigation, but individual and separate constructions in the surroundings of golf courses are indeed demanding of non-recycled water. With regard to this extensive model, it should be kept in mind that the buildings we are referring to are, to a large extent, second homes. The high-water demand of these residences comes from the maintenance of gardens and the filling of swimming pools. It is common to use tap water for garden maintenance and in many cases also for filling swimming pools. No data are available on the water used for this purpose, but there are data on the number of swimming pools in Spain. Newtral (2023), using the Dirección General del Catastro as a source, indicates that in Spain (not including the Basque Country and Navarra) there are 1.29 million swimming pools, of which it is estimated that 1.1 million are private. It does not have data on how many of these are in hotel establishments and how many in private residences, but obviously most of these are in the latter category. These data show that the demand for water for swimming pools is mainly originated in the private sector.

Considerations, Conclusions, and Recommendations

Tourism, like any economic activity, consumes water resources and must apply efficiency policies both to reduce costs and to improve its reputation and contribute to the sustainability of the system. Tourism is often criticized for its water consumption, especially in activities such as golf, which are perceived as elitist. However, tourism also brings social welfare and economic development, so it must make commitments at the same level as other sectors. Furthermore, the urban planning model is a determining factor in water consumption: a concentrated model, such as that of Benidorm, with high density and verticality, is more efficient than an expansive and dispersed one. Therefore, it is essential to rethink urban planning in tourist areas to reduce water impact.

On the other hand, both tourism and second homes should be fully integrated into water saving measures during episodes of water stress. This implies restrictions on watering gardens, filling swimming pools and non-essential use in hotels, limiting consumption to basic use. Although progress has been made in efficiency in destinations with concentrated models, the extensive model still has ample room for improvement. In this context, it is proposed to apply a progressive and dynamic tariff to water supply. Progressivity, similar to income tax, would allow basic consumption to have the lowest price, penalizing excessive consumption coming from ornamental irrigation or pool filling. The dynamic pricing system would automatically adjust the tariff scale according to the level of water stress, encouraging a more responsible use of the resource without affecting access to essential consumption.

References

Alicante Plaza (2025) La primera fase de la red de agua regenerada de Benidorm será en Poniente y llegará a 6.300 hogares [The first phase of Benidorm's reclaimed water network will be in Poniente and will reach 6,300 homes]. https://alicanteplaza.es/alicanteplaza/marinabaixa/la-primera-fase-de-la-red-de-agua-regenerada-de-benidorm-comenzara-en-la-nueva-zona-de-poniente#:~:text=Consumo%20en%20Benidorm,piscinas%20demandan%20otro%20medio%20hect%C3%B3metro (accessed June 30 2025)

Castro P, López J, Polaroo Analytics (2024) El consumo de agua en apartamentos y viviendas turísticas en España: Informe 2023. Polaroo. https://polaroo.com/reports/consumo-de-agua-en-sector-alquiler-vacacional-en-espana

EFE (2025) España podría llegar a los 100 millones de turistas extranjeros en 2025 pero no arrebatará el liderato a Francia [Spain could reach 100 million foreign tourists in 2025 but will not take the lead from France]

Gabarda Mallorquí A, Ribas Palom A, Daunis-i-Estadella J (2015) Desarrollo turístico y gestión eficiente del agua. Una oportunidad para el turismo sostenible en la Costa Brava (Girona). Investig Turísticas 9:50–69

Gallardo F (2025) El modelo de negocio del turismo en España: presente y futuro. Rev Economistas 188:109–118

Hosteltur (2023) Benidorm: cómo ser la tercera ciudad más eficiente del mundo en uso de agua [Benidorm: how to become the third most efficient city in the world in terms of water use] https://www.hosteltur.com/160699_benidorm-como-ser-la-tercera-ciudad-mas-eficiente-del-mundo-en-uso-de-agua.html (accessed June 30, 2025)

INE (Instituto Nacional de Estadística) (2025a) Movimientos turísticos en frontera. https://www.ine.es/jaxiT3/Tabla.htm?t=10822 (accessed May 30 2025)

INE (2025b) Viajes, pernoctaciones, duración media y gasto por comunidad autónoma de residencia de los viajeros. https://www.ine.es/jaxiT3/Datos.htm?t=12448#_tabs-tabla (accessed May 30 2025)

INE (2025c) Estadística sobre el Suministro y Saneamiento del Agua Año 2022. https://www.ine.es/dyngs/Prensa/ESSA2022.htm (accessed May 30 2025)

INE (2025d) Estadística experimental. Medición del número de viviendas turísticas en España y su capacidad. https://www.ine.es/experimental/viv_turistica/experimental_viv_turistica.htm (accessed May 30 2025)
INE (2025e) Resultados por comunidades autónomas. Serie 2000-2022. Suministro y saneamiento de agua. https://www.ine.es/dynt3/inebase/index.htm?padre=8709&capsel=8710 (accessed May 30 2025)
INE (2025f) Coyuntura Turística Hotelera (EOH/IPH/IRSH). Abril 2025. Datos provisionales. https://www.ine.es/dyngs/INEbase/operacion.htm?c=Estadistica_C&cid=1254736177015&menu=ultiDatos&idp=1254735576863 (accessed May 30 2025)
INE (2024) Estadística sobre el Suministro y Saneamiento del Agua - Año 2022. Nota de prensa. https://www.ine.es/dyngs/Prensa/ESSA2022.htm (accessed May 30 2025)
Kuzma, S, Bierkens, M, Lakshman, S, Luo, Saccoccia, L, Sutanudjaja, E, Beek, R (2023) Aqueduct 4.0: updated decision-relevant global water risk indicators. World Resources Institute, Washington, DC
López JA (2024) Golf y Agua: Informe sobre los campos de golf de España y su relación con el agua [Golf and Water: Report on Spain's golf courses and their relationship with water]. Universidad de Cádiz. https://www.aecg.es/wp-content/uploads/INFORME-SOBRE-LOS-CAMPOS-DE-GOLF-ESPANOLES-Y-SU-RELACION-CON-EL-AGUA_.pdf [accessed 1 July 2025]
MITECO (Ministerio para la Transición Ecológica y el Reto Demográfico) (2022) Informe de seguimiento de los planes hidrológicos de cuenca y de los recursos hídricos en España – Año 2021. Gobierno de España. https://www.miteco.gob.es/
Newtral (2023) España tiene una piscina por cada 35 personas: consulta cuántas hay en tu municipio [Spain has one swimming pool for every 35 people: find out how many there are in your municipality]. https://www.newtral.es/piscinas-habitante-espana/20230801/ (accessed June 30, 2025)
Olcina Cantos J, Baños Castiñeira CJ, Rico Amorós AM (2016) Medidas de adaptación al riesgo de sequía en el sector hotelero de Benidorm. Rev Geogr Norte Grande 65:129–153
Statista (2025) Porcentaje del PIB aportado por el sector turístico en España de 2010 a 2024 [Percentage of GDP contributed by the tourism sector in Spain from 2010 to 2024]. https://es.statista.com/estadisticas/1082929/sector-turistico-porcentaje-del-pib-aportado-espana/ (accessed June 30 2025)

Chapter 13
Inherited Desertification: Mining, Deforestation and Abandonment in the Sierra de Gádor Mountains (Almería)

María E. Sanjuán and Gabriel del Barrio

Abstract Inherited desertification refers to the state of a landscape after its past desertification and ecological degradation, which, although no longer active, have left an aftermath impeding any significant recovery. It is often associated with a global change event whereby significant socio-economic changes bring ephemeral wealth about at the expense of overexploiting a natural resource beyond its resilience. In Southeast Spain, in the Sierra de Gador, in the nineteenth century, this happened during the lead mining rush. The mountain population, pre-eminently agricultural, ventured in the enterprise of mining with chaotic manual labor. Lead made up over 80% of the total Spanish mineral/metallurgical exports until the 1870s. Smelting was carried out using the local woodlands, and the impact was irreversible. It is estimated that 1.4 million tons of alpha grass and over 52,000 tons of holm-oak charcoal had been consumed by 1860. These figures would be the equivalent of about 50,000 hectares of grasslands and around 28,000 hectares of holm-oak woods. Climate drift toward increased aridity and intense soil erosion caused by runoff over deforested slopes make recovery of these forests almost impossible at a human time scale. This sequence of processes stresses the most dramatic aspect of desertification: the enormous difference between the duration of its causes and that of its effects.

Keywords Desertification · Land degradation · Sustainability · Mining industry · Little Ice Age

M. E. Sanjuán (✉) · G. del Barrio
Estación Experimental de Zonas Áridas (EEZA), CSIC, Almería, Spain
e-mail: marieta@eeza.csic.es

G. del Barrio
e-mail: gabriel@eeza.csic.es

J. Martínez-Valderrama and J. Olcina Cantos (eds.), *The Labyrinth of Desertification*,
https://doi.org/10.1007/978-3-032-09648-7_13

Inherited Desertification, the Footprint of Excess

The most dramatic aspect of desertification is the enormous difference between the duration of its causes and that of its effects. Desertification in drylands occurs when a socioeconomic system exploits a natural resource beyond its threshold for recovery. This means a few decades of wealth are acquired at the cost of a natural capital accumulated over the ages, while the subsequent land degradation lasts centuries, an order of magnitude longer.

Inherited desertification refers to the state of a landscape after its past desertification and ecological degradation, which, although no longer active, have left an aftermath impeding any significant recovery.

The Sierra de Gador Mountain Range is an example of a landscape trapped by the footprint of past human activity. It is a rocky mass over 2200 m.a.s.l. funneling cyclonic west winds, and making the southeast Iberian Peninsula the most arid zone in Europe. Its relief is covered by alpha grass ("esparto"), gorse, pine reforestation, holm-oak woods that hardly merit the name, and high-altitude thorn bushes. In general, they are serial formations leaving the spotty original vegetation scattered and inaccessible.

The current socioeconomic scenario is in some ways parallel to the landscape. The main drivers orienting the economy since the end of the twentieth century have been rural tourism and real estate speculation (Collantes Gutierrez 2007). Such drivers

Fig. 13.1 Deforested slopes colonized by low secondary vegetation. The coastal urban nuclei polarize the new economic activity around tourism and greenhouses. Loma del Morrón del Aljíbe (Enix, Almería, Spain). Photo by María E. Sanjuán

operate easily in a rural population with a need to diversify its income, weighted down by the abandonment of traditional farming in a natural setting already exhausted by mining. The younger population has seen the new agriculture under plastic on the nearby coast as their opportunity, and this, in turn, has led to further desertification, not discussed here (Fig. 13.1). Thus, a new stage was set in a society already clearly in recession since the end of the nineteenth century.

Mining in the Sierra de Gador Mountains

During the first half of the last century, the population bottomed out with unprecedented emigration. The reasons were the mining crisis and the decline in cultivation of the "boat grape" (Marin Corbera 2010). Both activities modified the provincial landscape. But it was mining that destroyed the natural heritage of the Sierra de Gador Mountains.

Its wealth in minerals was known in the remote past, but mass exploitation did not occur until the nineteenth century. The Sierra de Gador Mountains became a reference in Spanish mining in 1817. Lead made up over 80% of the total Spanish mineral/metallurgical exports until the 1870s (Perez de Perceval Varde and Sanchez Picon 2001).

Circumstances were right: increased international demand, sector privatization, the mountains' distinctive geological characteristics and strategic location, and the concurrent socioeconomic structure. All of this together led to a spectacular mining fever. With the wave of generalized optimism of the first industrial revolution in the background a false perception of unlimited growth arose, in which "nature began to be conceived as simply an inert and inexhaustible depository of resources" (Garcia Latorre and Garcia Latorre 2007).

The Spanish Parliament's (*Cortes del Trienio Liberal*) decree of October 25, 1820, stating that "...private individuals may benefit from any type of mine," was the beginning of large-scale mining. The mountain population, preeminently agricultural, saw a way to supplement their income, and at first, ventured in the enterprise with chaotic manual labor (Fig. 13.2). This involved hardly any extraction costs, since the Gador Mountains' natural drainage and the scant depth worked at avoided costly drainage. In addition, the hardness and fissures of the limestone minimized the need for shoring and natural ventilation. Small smelting ovens called "boliches" prospered in the gullies. These were later replaced first by shaft ovens and then by coke ovens. One aspect crucial to marketing was the location of the Sierra de Gador close to the coast, especially the port of Adra. This proximity facilitated transport by mule train without the need for mechanized transport, which would have been complicated by the rough relief (Perez de Perceval Verde 1989).

Mining led to a multitude of direct and indirect jobs and was a focal point of population up to 1840. It also led to an increase in capital, which made agricultural expansion for local subsistence possible under expropriation laws, and opened new horizons for export which took advantage of contacts made in exporting minerals. Vineyards producing "boat grapes" were one of the consequences, especially in the second half of the nineteenth century (Perez de Perceval Verde 1984).

Fig. 13.2 Abandoned manual mines are found throughout the Sierra de Gador. Paraje de los Arances (Almería). Photo by Jaime Martínez Valderrama

The environmental impact on the holm-oak forests in the Sierra de Gador (Garcia Latorre and Garcia Latorre 2007) was irreversible. Firewood for cooking in such a numerous population punished hillside vegetation, but smelting razed the landscape. The early boliches used for smelting were responsible for burning humongous amounts of alpha grass. Clumps were usually pulled up by the roots, even though it was prohibited by law, causing anxiety among esparto craftsmen, who could see how the fields were diminishing. The shaft ovens made use of the slag unused in the boliches in a second smelting. These required a source of greater heat and burned mainly holm-oak wood (Fig. 13.3).

It is estimated that 1.4 million tons of alpha grass and over 52,000 tons of holm-oak charcoal had been consumed by 1860. These figures would be the equivalent of about 50,000 hectares of esparto grasslands and over half a million trees, corresponding to around 28,000 hectares of holm-oak woods (Sanchez Picon 1998). For a better idea of the extent of this devastation, the entire Turón-Sierra de Gador geographic macrosystem occupies about 70,000 hectares (Oyonarte et al. 2016).

Manual mining began to decline as the forest resources it depended on were consumed. At the same time, lead ore processing with coal began to be industrialized along the coast. Prices began to decline starting in 1837, partly governed by the large smelting companies, and this was the end of manual mining due to its weak capitalization (Perez de Perceval Verde 1984). The natural resource had been used up and the

Fig. 13.3 Holm-oaks can only germinate in the shade of other trees, and therefore, this solitary tree is testimony to the holm-oak woodland that once extended throughout the Sierra de Gador. Paraje de la Zarba (Enix, Almería, Spain). Photo by María E. Sanjuán

window for financial opportunity had closed. And so, this episode of desertification concluded after a few decades of relative prosperity almost as quickly as it had begun.

Irreversible Environmental Impact

The land degradation derived from that episode is still visible over a century later. One of the immediate consequences of deforestation was intense runoff, causing heavy flooding downstream. This stimulated reforestation policies from the beginning of the twentieth century, with varying success.

There is an essential reason this degradation has been so long-lasting: the original soil has practically disappeared, eroded by the increase in surface runoff. The most frequent type of soil in the Gador Mountains were the rendzinas, shallow, well-structured humus-rich soils that form over calcareous bedrock under certain moisture conditions (Fig. 13.4). Those soils maintained positive feedback with the forest cover. After erosion, skeletal soils on predominant bedrock or lithic fragments retaining some soil remain (Li et al. 2007), but they are insufficient for non-pioneering species to grow. A 40–50-cm thick layer of rendzina requires several thousand years to form, and a more humid climate than the present one.

Fig. 13.4 Residual rendzina in cracks in dolomite lithosols in the Sierra de Gador. This discontinuous eroded "soil" is an average <10-cm thick, alternating with rocky outcrops. Xerophytic plants in the area develop in soil in the cracks. Photo by Albert Solé Benet

The age of the surviving trees is compatible with the incidence of the Little Ice Age (LIA), a cold period in the European region, which in the Iberian Peninsula extended from the sixteenthth to the nineteenth century and consisted of a succession of extreme precipitation and drought events (Martin Vide and Olcina Cantos 2001). The LIA meant a significant increase in precipitation in southern Iberia (Sousa and García-Murillo 2003), which could have fostered the development of the existing holm oak forests and even the regrowth of stumps from trees felled by clear-cutting during the mining heyday. The LIA ended in the mid-nineteenth century, and since then, the climate has been aridifying. Already established forests could have survived the new conditions thanks to their capacity for creating their own microclimate and the depth of their roots. But the current climate in the Sierra de Gador hardly enables the development of true forest cover, even if the soil were suitable (Fig. 13.5).

The ecological history of the Sierra de Gador is repeated in many other scenarios throughout Spain, in what could be termed global change events. Some examples are tilling to expand dryland cereals after the expulsion of the Moors (sixteenth–seventeenth centuries); successive "desamortizacion" consisting of the expropriation and liquidation of forest lands to amortize the national public debt (eighteenth–nineteenth centuries); or membership in the European Economic Community, which initiated a trend to livestock farming, intensive agriculture and gradual abandonment of marginal lands (twentieth century).

Fig. 13.5 Several new holm-oak boles resprouting from the stump after the main tree was cut down during nineteenth century deforestation. Neither the soil surrounding it nor the current climate could favor germination of an acorn. Photo by María E. Sanjuán

Inherited desertification involves an irreversible change in the state of the landscape toward a simple, banal configuration with no resources to face future uncertainties. They are systems that have lost their ecological capital, and with it, their options for management (del Barrio et al. 2021). All of this is occurring under transitory climatic conditions whose evolution is already impossible to keep up with. However, the most terrifying part of inherited desertification is that in their day, nobody cared about the trajectory undertaken.

Acknowledgments This research was financially supported by the project "Plan Complementario de I+D+i en el área de Biodiversidad (PCBIO)" funded by the European Union within the framework of the Recovery, Transformation and Resilience Plan - NextGenerationEU and by the Regional Government of Andalucia.

References

Collantes Gutierrez F (2007) La desagrarización de la sociedad rural española. Hist Agrar 42: 251–276. 1950-1991

del Barrio G, Sanjuán ME, Martínez-Valderrama J et al (2021) Land degradation means a loss of management options. J Arid Environ 189:104502. https://doi.org/10.1016/j.jaridenv.2021.104502

Garcia Latorre J, Garcia Latorre J (2007) Almería, hecha a mano una historia ecológica. Fundacion Cajamar, Almeria

Li XY, Contreras S, Solé-Benet A (2007) Spatial distribution of rock fragments in dolines: a case study in a semiarid Mediterranean mountain-range (Sierra de Gádor, SE Spain). *Catena* 70:366–374. https://doi.org/10.1016/j.catena.2006.11.003

Marin Corbera M (2010) Una tradición forjada a partir de 1939. Andalucia Hist 28:12–28

Martin Vide J, Olcina Cantos J (2001) Climas y tiempos de España. Alianza Editorial, Madrid

Oyonarte C, Giménez E, Villalobos M, Guirado J (2016) Sierra de Gádor, patrimonio natural e infraestructura verde de Almería. Fundación Patrimonio Natural, Biodiversidad y Cambio Global, Almeria

Perez de Perceval Varde MA, Sanchez Picon A (2001) El plomo en la mineria española del siglo XIX. Evolucion del sector y panorama empresarial. Fundación Empresa Publica, Madrid

Perez de Perceval Verde MA (1989) La mineria almeriense contemporanea (1800–1930). Zejel, Almeria

Perez de Perceval Verde MA (1984) Fundidores, mineros y comerciantes. La metalurgia de Sierra de Gador, 1820–1850. Cajal, Almeria

Sanchez Picon A (1998) Impacto sobre el medio ambiente de la industria metalurgica del siglo XIX. El Eco Alhama 5:1–4

Sousa A, García-Murillo P (2003) Changes in the wetlands of Andalusia (Doñana Natural Park, SW Spain) at the end of the Little Ice Age. Clim Change 58:193–217. https://doi.org/10.1023/A:1023421202961

Chapter 14
Irrigation as a Source of Development and Degradation

Julia Martínez Fernández

Abstract The idea that irrigation is a brake on the advance of the desert is often heard, based on the erroneous association between aridity and desertification and the attraction for green landscapes. However, in parallel with the decline of traditional orchards, new highly intensive agro-industrial irrigation has been expanding for decades, causing various environmental impacts, many of them associated with desertification processes. These effects include overexploitation of aquifers, which in turn has led to the loss or degradation of springs and associated ecosystems, wetlands and other water points. This increasing aridification of the landscape must be interpreted as a desertification syndrome linked to the unsustainable use of water. The expansion of agro-industrial irrigation has led to other effects associated to desertification, such as the salinization of water and soils and the increase in erosion processes, along with other negative impacts as the occupation of natural habitats, the increase in diffuse agricultural pollution, the generation of eutrophication processes as the one affecting the Mar Menor coastal lagoon (Murcia, Spain) and the increased flood damage due to increased runoff and sediment trawling from these agro-industrial irrigations. A just water transition is required to reduce desertification associated with poor water management.

Keywords Irrigation · Desertification · Impacts · Overexploitation · Just water transition · Southeastern Spain · Segura basin

J. Martínez Fernández (✉)
Fundación Nueva Cultura del Agua, Zaragoza, Spain
e-mail: julia@fnca.eu

J. Martínez-Valderrama and J. Olcina Cantos (eds.), *The Labyrinth of Desertification*,
https://doi.org/10.1007/978-3-032-09648-7_14

Irrigation: A Brake on the Advance of the Desert or an Engine of Desertification? The Case of the Peninsular Southeast

The idea that irrigation is a brake on the advance of the desert is often heard. This idea is based on the frequent but erroneous association between aridity and desertification and the attraction for green landscapes. In addition, irrigation is much more productive than rainfed crops. In fact, irrigation in Spain occupies 22% of the total agricultural area (MAPAMA 2023) but generates two-thirds of agricultural employment and agricultural added value (FNCA 2019). However, irrigation is not a homogeneous reality and covers very different situations, from historical and traditional irrigation to intensive and agro-industrial.

Historical irrigation represents 25% of Spanish irrigation (FNCA 2019) and constitutes agroecosystems of high environmental and cultural value (Hermosilla 2010). One of the reasons for their durability over time and environmental value is that they are located in the vicinity of springs and rivers, which provide renewable water and soils of high natural fertility (Fig. 14.1). However, these traditional orchard agro-landscapes are disappearing due to their lower profitability in current markets and their transformation to urban uses (Martínez Fernández et al. 2013; Martínez-Paz et al. 2019).

In parallel with the decline of traditional orchards, new highly intensive agro-industrial irrigation has been expanding for decades, often in the hands of large companies

Fig. 14.1 Traditional garden of the Segura River in the region of Valle de Ricote (Murcia). Photo by Julia Martínez Fernández

Fig. 14.2 Agroindustrial irrigation in Campo de Cartagena (Murcia). Photo by Julia Martínez Fernández

(Calvo Tornel 2006; Soler and Fernández 2022) that use precarious labour (Justicia Alimentaria 2020; Martínez-Valderrama et al. 2024). These irrigations do not require the proximity of rivers or springs or fertile soils, because their highest productivity is based on the use of a large number of inputs, including water (which can come from distant sources or from overexploited aquifers), synthetic fertilizers, pesticides or plastics (Fig. 14.4). These irrigations generate various environmental impacts, many of them associated with desertification processes. This is the case of agro-industrial irrigation in the Iberian Southeast (Martínez-Fernández et al. 2021; Zuluaga Guerra et al. 2023; Martínez-Valderrama et al. 2024) (Fig 14.2).

Irrigation and Desertification: Scaling Up

Irrigation, very productive and dominated by the color green, can be perceived as alien to desertification, even as a barrier to it, when in fact they can be active engines of desertification. To this perception contributes the fact that the desertification generated is not so evident, because it affects non-visible systems such as aquifers or takes place in a space different from that occupied by irrigation. That is why it is necessary to expand the scale and consider a wider territory that covers the abstraction sources, the irrigated perimeters and the receiving systems of the excess irrigation, usually generators of contamination by fertilizers, pesticides and salts.

A good example is the irrigation of the Segura basin, within the Iberian Southeast, where throughout the last century various hydraulic projects were carried out (first reservoirs, the Tagus-Segura transfer since the 1980s) to eliminate the water deficit and expand the irrigated area. However, these projects not only did not eliminate the deficit but also contributed to its increase. The expectations created by these hydraulic projects, especially by the Tagus-Segura transfer, encouraged the expansion of irrigation well beyond the available resources, contributing to generate first and then increase the water deficit, leading to a spiral of unsustainability (Martínez Fernández et al. 2021), in which the supposed solution (increase in the supply of water resources) only aggravates the problem (Zuluaga Guerra et al. 2023).

The need to address irrigation well above available resources has led to increasing overexploitation of aquifers, which in turn has led to the loss or degradation of various water-linked ecosystems, in particular fontinal sources and ecosystems (Fig. 14.3), wetlands and other water points. This increasing *aridification of the landscape* must be interpreted as a desertification syndrome linked to the unsustainable use of water (Fig. 14.4).

Along with the overexploitation of water and the aridification of the landscape, the expansion of agro-industrial irrigation has led to other effects in the Segura basin, including the salinization of water and soils (Fig. 14.5), another process ascribed to desertification; occupation of natural habitats (Esteve et al. 1995; Caballero et al. 1996); the increase in diffuse agricultural pollution, highlighting the eutrophic crisis in the coastal lagoon of the Mar Menor due to the massive inflow of nutrients (nitrogen and phosphorus) from the Campo de Cartagena (Fig. 14.5, Esteve Selma et al. 2024); the increase of erosion (another process associated with desertification) by the tillage of inadequate soils and by transformation to very intensive irrigation without vestiges of natural vegetation (Fig. 14.6) or conservation practices (Martínez-Valderrama et al. 2020) and increased flood damage due to increased runoff and sediment transport from these agro-industrial irrigations.

Desertification syndromes due to unsustainable water use are not confined to the Iberian Southeast. In Spain, irrigation demand accounts for 80% of the water used (MITERD 2021) and more than 90% of net consumption (Albiac et al 2023). Its continued expansion is one of the main pressures on the quality, productivity and

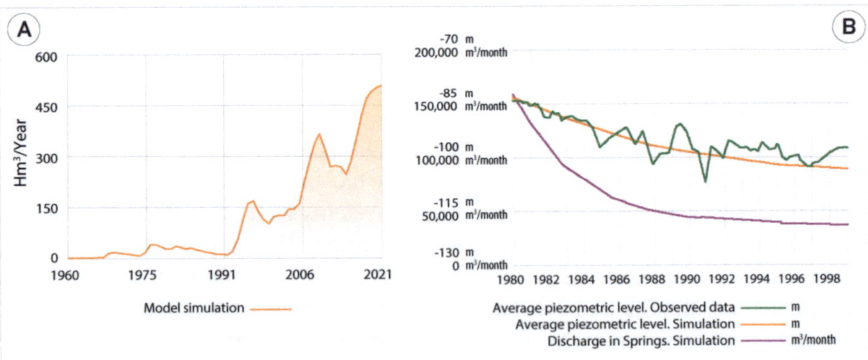

Fig. 14.3 **A** Evolution of the overexploitation of groundwater in the Segura basin. Results of the simulation of the 'Cuenca del Segura' model. Source: Zuluaga Guerra et al. (2023). **B** Evolution of the average piezometric level and the loss of the discharge of the springs in the area of Mazarrón-Águilas (Segura basin) due to the overexploitation of aquifers. In orange and pink the results of the simulation of the 'Intensive Irrigation' model and in green the observed values. Source: Martínez Fernández and Esteve Selma (2002) and Martínez Fernández et al. (2021)

biodiversity of ecosystems and on water bodies, half of which do not reach Good Status due to excessive river abstractions, overexploitation of aquifers and contamination by fertilizers and pesticides, among other factors (FNCA 2023). Proof of this is the reduction of river flows, the nitrate pollution of many aquifers, the alarming loss of wetlands, affecting protected areas of international relevance such as the Tablas de Daimiel or Doñana (Client Earth 2024; Sanchís Ibor and Ibáñez Martí 2024) and the increasing loss of aquatic biodiversity.

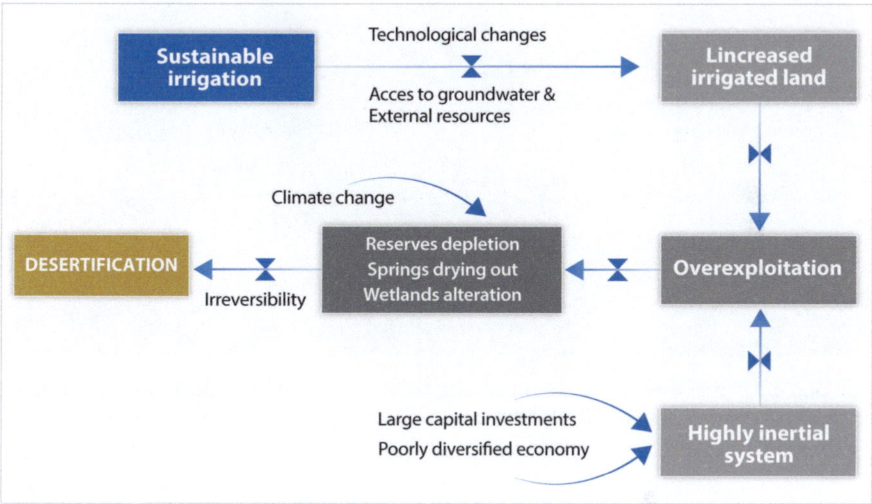

Fig. 14.4 Outline of the desertification process induced by the unsustainable use of irrigation in the Segura basin. Source: Martínez Fernández et al. (2021)

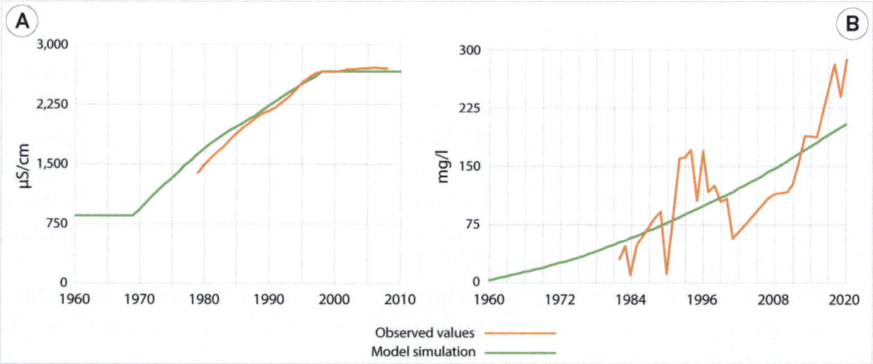

Fig. 14.5 **A** Increase in salinity in the Segura River caused mainly by the expansion of irrigation. **B** Evolution of nitrate content in the groundwater of Campo de Cartagena. Source: Zuluaga Guerra et al. (2023)

Fig. 14.6 Soils are completely devoid of vegetation in the intensive irrigation of Campo de Cartagena (Murcia). Photo by Julia Martínez Fernández

Reorienting Irrigation in Light of a Just Water Transition

In response to the great pressure of irrigation on water bodies and ecosystems, different measures have been proposed. One of the most applied and demanded is the modernization of irrigation, which is a false solution because, far from saving water, in many cases it increases its consumption by reducing the irrigation returns that feed rivers and aquifers and by intensifying production, which implies greater evapotranspiration and therefore greater net water consumption (Grafton et al. 2018; Aliod and Fernández 2023; Pérez-Blanco et al. 2023). In addition, gravity irrigation currently represents less than 20%, so that localized, sprinkler and automotive irrigation account for more than 80% of the total, highlighting the localized one because it is the majority and the one that grows the most (MAPAMA 2023), without this leading to a lower pressure on rivers and aquifers (Food Justice 2020; Pérez-Blanco et al. 2021).

Reducing desertification associated with poor water management requires a change of approach, a just water transition with the triple objective of adapting to climate change, guaranteeing human supply and restoring the good status of our water bodies and ecosystems. A key element for this transition is the reduction of the irrigated area, with the participation of all stakeholders and through a social distribution of water, in order to reduce part of the agro-industrial irrigation, often in the hands of large companies, while protecting historical and traditional irrigation, small farmers and family farming. Along with this reduction, the de-intensification and environmental conversion of agro-industrial irrigation is necessary in order to guarantee the present and future of agricultural landscapes, ecosystems and the entire territory in the face of the challenges of climate change.

References

Albiac Murillo J, Esteban García E, Baccour S (2023) The Situation and Prospects of Water Resources in Spain. Studies on the Spanish economy. FEDEA. Fundación de Estudios de Economía Aplicada. Online document. https://documentos.fedea.net/pubs/eee/2023/eee2023-29.pdf

Aliod R, Fernandez A (2023) New indications of the increase in water and energy consumption after irrigation modernization programs in Spain. XII Iberian Congress of Water Management and Planning. Looking at the rivers from the sea: old and new debates for a just water transition. University of Murcia, p 628–635. https://doi.org/10.6018/editum.3003

Caballero JM, Calvo JP, Esteve MA, Nicolás E, Robledano F (1996) Saladares del Guadalentín, Spain. In Management of Mediterranean. Wetlands 2:109–127

Calvo García-Tornel F (2006) Spanish Southeast: irrigation, hydraulic technologies and territorial changes. Scripta Nova 10:218

Client Earth (2024) Report on Spanish third cycle hydrological plans: consideration of climate change and key aspects in the implementation of the Water Framework Directive. Client Earth. Online publication. https://www.clientearth.es/media/pk0hrat1/executive-summary-english.pdf

Esteve MA, Caballero JM, Giménez A et al (1995) The landscapes of the water in the region of Murcia. Environmental characterization and management perspectives of wetlands. In: Water and future in the Region of Murcia. Regional Assembly of Murcia, Murcia, p 301–341

Esteve Selma MA, Giménez Casalduero F, Martínez-Fernández J, Robledano Aymerich F (2024) Structural reasons for the collapse of the Mar Menor from a socio-ecological approach. In: Ibor S, Martí I (eds) The coastal wetlands of the Iberian Peninsula. Editorial Tirant Humanities, Valencia, p 111–153

FNCA (2019) Mediterranean agriculture and water. Some keys. Working Group "Mediterranean Agriculture and Water". New Water Culture Foundation. https://fnca.eu/library-of-water/directory/file/2899-agriculture-mediterranean-and-water-some-keys?start=20

FNCA (2023) Overall assessment of the third cycle management plans. OPPA Report 2022. New Water Culture Foundation. Online document. https://fnca.eu/biblioteca-del-agua/documentos/documentos/PLANES_OPPA22.pdf

Grafton RQ, Williams J, Perry CJ et al (2018) The paradox of irrigation efficiency: higher efficiency rarely reduces water consumption. Science 361:748–750

Hermosilla J (2010) Historical Spanish irrigation: cultural landscapes, sustainable landscapes. Ministry of the Environment and Rural and Marine Affairs, Madrid

Food Justice (2020) The secret ingredient. Labour exploitation in Spanish food. Justicia Alimentaria. Online report. https://justiciaalimentaria.org/wp-content/uploads/2022/10/ingredientesecreto.pdf

Martínez Fernández J, Esteve Selma MA (2002) Water, irrigation and sustainability in the Iberian Southeast. Bakeaz-New Water Culture Foundation, Bilbao

Martínez-Fernández J, Esteve-Selma MA, Banos-González I, Carreño F, Moreno A (2013) Sustainability of Mediterranean irrigated agrolandscapes. Ecol Model 248:11–19

Martínez-Fernández J, Esteve Selma MA, Zuluaga Guerra PA (2021) Water and sustainability. Towards a water transition in the Iberian Southeast. Ecosistemas 30:2254

Martínez-Paz JM, Baños-González I, Martínez-Fernández J, Esteve-Selma MA (2019) Assessment of management measures for the conservation of traditional irrigated lands: The case of the Huerta of Murcia (Spain). Land use policy 81:382–391
Martínez-Valderrama J, Guirado E, Maestre FT (2020) Unraveling misunderstandings about desertification: the paradoxical case of the Tabernas-Sorbas Basin in Southeast Spain. Land 9:269
Martínez-Valderrama J, Gartzia R, Olcina J, Guirado E, Ibáñez J, Master T (2024) Uberizing agriculture in drylands: a few enriched, everyone endangered. Water Resour Res 38:193–214
MAPAMA (2023) Survey on areas and crop yields in Spain. ESYRCE. Ministry of Agriculture, Fisheries and Food, Spain
MITERD (2021) Monitoring report of hydrological plans and water resources in Spain. Ministry for Ecological Transition and the Demographic Challenge, Spain
Pérez-Blanco D, Loch A, Ward F, Perry C, Adamson D (2021) Agricultural water saving through technologies: Zombie idea. Environ Res Lett 16:114032
Sánchez Ibor C, Ibáñez Martí C (2024) The coastal wetlands of the Iberian Peninsula. Editorial Tirant Lo Blanch, Valencia
Soler C, Fernandez F (2022) Structure of land ownership in Spain. Concentration and hoarding. Mundubat Foundation and Food Sovereignty, Biodiversity and Cultures Magazine. Online report. https://www.mundubat.org/informe-mundubat-acaparamiento-de-tierras-en-espana-2016/
Zuluaga-Guerra PA, Martinez-Fernandez J, Esteve-Selma MA, Dell'Angelo J (2023) A socio-ecological model of the Segura River basin, Spain. Ecol Model 478:110284

Chapter 15
Desertification in Olive Orchards: A Complex Challenge for a Diverse Agricultural System

José Alfonso Gómez

Abstract This chapter examines desertification risk within Spanish olive cultivation, a dominant Mediterranean agricultural system. It defines four main orchard typologies: traditional (steep slopes/flat-rolling landscapes), intensive, and super-intensive, highlighting a significant productivity gradient between them. The primary sustainability threats identified are soil degradation (primarily water erosion and low organic carbon) and pressure on water resources due to irrigation expansion. Historically, soil management practices maintaining bare soil accelerated erosion, with studies showing losses quadrupling (e.g., from ~20 to ~80 t ha^{-1} $year^{-1}$) post-1960s mechanization. Research demonstrates that temporary cover crops (CV) significantly reduce erosion and improve topsoil quality compared to tillage (LC) or no-till herbicide (NLH) management. However, CV implementation carries a significant risk of yield loss due to water competition if not managed properly. Irrigation, largely deficit-based, has increased yields but intensifies pressure on water resources, especially in vulnerable areas, potentially worsening with climate change. Despite this, economic constraints on farms and technical advisory gaps hinder widespread adoption of conservation practices. Addressing desertification requires integrated farm-level solutions: optimizing cover crop use, enforcing water governance, supporting sustainable rainfed production, compensating land-use reversion, and developing drought-resistant varieties. The diversity of olive systems requires tailored strategies combining socioeconomic and technical measures.

Keywords Woody crops · Mediterranean · Soil erosion · Water resources · Soil quality

---

J. Alfonso Gómez (✉)
Instituto de Agricultura Sostenible (IAS), CSIC, Córdoba, Spain
e-mail: joseagomez@ias.csic.es

J. Martínez-Valderrama and J. Olcina Cantos (eds.), *The Labyrinth of Desertification*,
https://doi.org/10.1007/978-3-032-09648-7_15

Spain's Agricultural Landscape: The Reign of the Olive Grove

Olive cultivation is an agricultural system widely distributed on both shores of the Mediterranean, where approximately 8 million hectares are cultivated, 2.8 million of them in Spain. It is also a crop deeply rooted in the territory and culture of many areas of Spain and other Mediterranean countries. This vast distribution, and its long and intense interweaving with the territory (Fig. 15.1), makes it prone to generalizations that obscure its reality. In this section, we will attempt to shed some light on this crop to better understand the problems it poses in relation to desertification and explore strategies to address it.

The majority of Spanish olive orchards are located in Andalusia, Castilla-La Mancha, Extremadura, Catalonia, and Aragon, accounting for approximately 60%, 16%, 11%, 4%, and 2% of the total area. Historically, olive orchards exploited different resources, for example, combining the production of olives (for oil or table olives) with grazing or intercropping of cereals or legumes, and the exploitation of pruning wood. However, since the late nineteenth century, olive orchards specialized in olive production (Infante-Amate et al. 2014). Despite this overall trend, there are some regions where agro-forestry-pastoral use is maintained, such as in mountain olive groves in the Sierra Morena mountain range in Southern Spain (Álvarez et al. 2007).

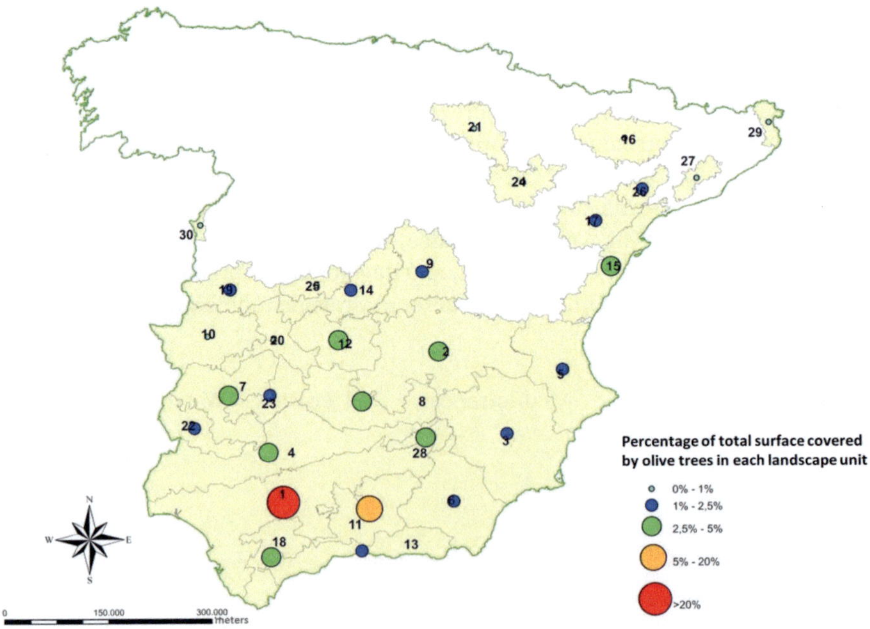

Fig. 15.1 Percentage of land occupation by olive groves in the different landscape units proposed for the Spanish peninsular olive grove by Guzmán et al. (2020). Source: Adapted from Hernández (2011)

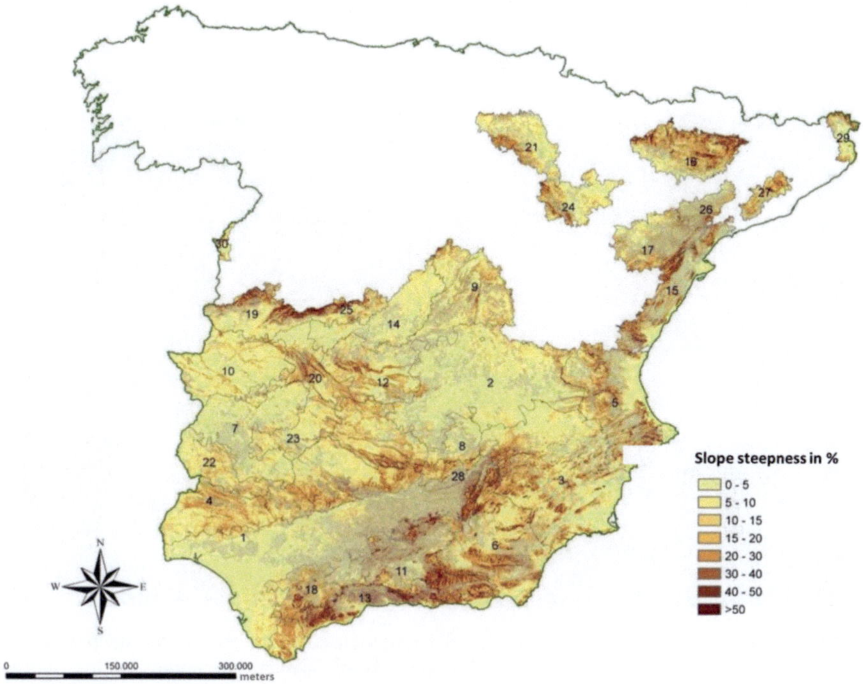

Fig. 15.2 Distribution of slope steepness in the Spanish peninsular olive orchards. Source: Adapted from Hernández (2011)

The current distribution of olive groves in Spain is the result of a combination of physiological, soil, and climate factors, and socioeconomic processes. As a highly hardy crop, highly adapted to Mediterranean conditions and adaptable to different levels of cultivation intensity, it is found across a very wide range of soils and slopes, e.g., Fig. 15.2.

This historical evolution has led to a situation in which different types of olive orchards coexist today (Infante-Amate et al. 2014; Guzmán et al. 2020). Figure 15.3 illustrates this diversity, which can be summarized in a simplified manner into four major typologies. On the one hand, there is the traditional olive grove on steep slopes (or mountain). This type of olive grove dates back to the expansion of olive groves to these areas before the twentieth century, and it is mostly rainfed. Most of these olive orchards present a low productivity and greater operating difficulties and cost due to their topography. The majority of these olive orchards were established without the introduction of soil conservation measures, although in some areas, such as the eastern part of the Iberian Peninsula, they were terraced. A second typology is the traditional rainfed olive orchard in flat or rolling landscapes. This type of olive orchard presents a higher productivity, because it is located on better soils. It also presents fewer operating limitations. Until the olive-growing revolution around the 1980s,

Fig. 15.3 Four major simplified olive orchard typologies. **A** Traditional on steep slopes. **B** Traditional on flat or rolling landscapes. **C** Intensive. **D** Super-intensive. Photos by José Alfonso Gómez

this was considered the most profitable kind of olive orchard. From that time on, thanks to agronomic improvements and mechanized cultivation, two other types of olive orchards emerged. Intensive olive groves emerged, using higher plant density, single-trunk trees, and, if available, deficit irrigation to achieve higher yields and economy efficiency. This type intensive olive orchard encompasses a combination of converted traditional plantations or new plantations located that have substituted old traditional plantations or new ones in farms formerly dedicated to other crops. Since the end of the twentieth century, a fourth type of olive orchard, the super-intensive, has become a growing reality. This system focuses on hedgerow cultivation at very high planting density to facilitate harvesting and accelerate the plantation's entry into production. These orchards are mostly deficit-irrigated, although it is a cultivation system adaptable to dryland conditions by modifying the planting density. Overall, there is a very pronounced gradient of productivity between these different types of orchards, increasing from A to D (see Fig. 15.3), with olive harvests of less than 500 kg ha^{-1} $year^{-1}$ in some traditional dryland olive groves on steep slopes up to harvests above 10–12,000 kg ha^{-1} $year^{-1}$ in super-intensive olive groves (Gómez 2009).

It is difficult to provide a comprehensive distribution among these four categories of olive orchards for the whole of Spain, but an estimate for Andalusia (60% of the surface area) suggests that 30% is traditional on steep slopes (steepness above

15%), 35% is traditional dryland, 27% is intensive or traditional irrigated and 8% is super-intensive. These figures should be viewed with caution in absolute terms, but they provide a reasonable approximation of the relative weight of different olive grove typologies in Spain.

This dynamic of olive cultivation in recent decades has been driven by a growing global demand for olive oil, technological improvements, a greater global integration of the olive oil and table olive markets and, in the case of European Union countries, a favorable regulatory and public aid framework within the Common Agricultural Policy. All of this has resulted in a mosaic of olive orchards that reflect the socioeconomic reality at the time of their establishment (Guzmán 2004).

The Sustainability of the Spanish Olive Cultivation in Relation to Desertification

The Spanish olive cultivation systems, or as a subrogate the Andalusian one which represent around 62 and 80% of the Spanish olive surface and yield, respectively, has been the subject of various analyses regarding its sustainability (e.g., Beaufoy 2001; Gómez 2009; Scheidel and Krausmann 2011). The main threats identified at the national level can be summarized into five main ones:

1. Soil degradation, mainly due to accelerated water erosion and low organic carbon content in the soil.
2. Reduction of biodiversity due to farming intensification at the exploitation and landscape simplification.
3. Pressure on water resources, due to the expansion of the irrigated surface.
4. Diffuse pollution due to nutrients and agrochemicals transported by runoff and sediment to surface water bodies and percolation to groundwater bodies.
5. Low economic viability of many farms, especially small and medium-sized ones, under less favorable production conditions.

Added to these are the appearance of emerging diseases or pests. Among these is the fungus that causes verticillium wilt (Montes-Osuna and Mercado-Blanco 2020), which causes significant damage to olive groves each year, although it does not threaten the overall sustainability of the system. Also affecting the bacterium *Xylella fastidiosa* (De la Fuente 2024), which is not yet a serious problem in Spain, but has already caused severe damage in southern Italy and remains a major threat.

The intensity and significance of the five threats mentioned above vary depending on the type of olive grove and its soil, climate, and socioeconomic conditions. Nevertheless, there are a number of general trends in relation to desertification that can be discussed and commented below, to provide a general understanding and explore strategies for addressing them. Soil degradation and pressure on water resources are the two threats to sustainability directly related to the risk of desertification. The first, soil degradation, will be discussed below, always in the context of its diversity, followed by a brief overview of the pressure on water resources.

Soil Degradation in Olive Groves

Olive orchards are characterized by being a crop with low ground cover per tree, because planting density and canopy size are adapted to ensure the viability of the plantation in conditions with limited water availability, and a soil management that has historically been based on eliminating all adventitious vegetation between trees to limit competition for water and nutrients with the olive tree (Vanwalleghem et al. 2011). These agronomic conditions, combined with cultivation on slopes (in many cases very steep and without terracing) and regularly intense rainfall events, suggest that this crop was subjected to accelerated water erosion processes before its intensification from the second half of the twentieth century. Historical erosion studies in olive groves are not very numerous, and must necessarily be cautious in their extrapolations, but some of those that exist seem to support this idea. Thus, Vanwalleghem et al. (2011) in a study on cumulative erosion rates in three olive groves planted in the 18th and 19th centuries in a mountainous area of Andalusia, reported an average loss of about 45 cm of soil until 2010. References to evident symptoms of erosion in olive orchards in the press of the early twentieth century (e.g., Rodrigáñez 1908), or in specialized literature (Bennet 1960), and an understanding of the dynamics of erosive processes suggest that problems of soil degradation due to erosion have been present in olive orchards for centuries, although possibly at a much lower intensity than that occurring since the intensification of practices in this crop since the 1960s, approximately.

From that date onwards, the widespread expansion of the agricultural tractor, and later of herbicides, as well as the abandonment of other uses of the olive grove (e.g., grazing, etc.), made it possible to achieve the objective of maintaining the soil bare throughout the year much more effectively. It was at this point, which can be dated, depending on the region, to sometime between the early 1960s to the mid-1970s, that an olive landscape dominated by bare soil was established. This transition towards accelerated erosion processes in many olive-growing areas is something documented in some historical erosion studies, such as the aforementioned Vanwalleghem et al. (2011). These authors, using simulation models, estimated that it was at this point that soil losses due to water erosion at the hillside scale approximately quadrupled, from an average of approximately 20 t ha^{-1} year^{-1} to 80 t ha^{-1} year^{-1}. It was at this time that management changes combined with landscape simplification through the elimination of natural vegetation, such as hedges, accelerated the formation of gullies in some olive grove areas, such as the inland areas of the Guadalquivir Valley, which rest on easily eroded sedimentary materials (Hayas et al. 2025). This was the period when some of the first technical work aimed at erosion control in olive groves began to appear, mainly the reintroduction of management using cover crops, which we will discuss later (Ruiz de Castroviejo 1969).

Since the late twentieth and early twenty-first centuries, there have been various experimental studies, although not numerous, that have measured water erosion in experimental plots at the hillside and small watershed scales using natural rainfall events. These studies have focused primarily on quantifying differences in soil loss and runoff between systems based on bare soil, either through tillage (TL) or no-till herbicide (NTH), versus others that use different types of cover crop or a mulch of chopped pruning residues in the alley. Cover crop in olive orchards (CV) is almost always a temporary cover crop that is left to grow in the alley during autumn-winter and is mechanically or chemically controlled at the end of winter to avoid competition for water and nutrients with the olive tree. Table 15.1 shows a summary of these studies, which show the differences between systems and how management with CV can significantly reduce soil loss compared to systems with bare soil. For convenience, these studies typically focus on these three types of soil management, although the reality on farms is much more complex. When the actual farm management are looked in detail, several studies describe numerous ways of applying these management methods (Gómez et al. 2021), using them individually (e.g., number of tillage operations, canopy extension, etc.) or in combination (e.g., mulching in the street and herbicide under the canopy), and varying in other features that are key to determine water erosion like tillage intensities, ground cover by the olive canopy, etc.

Table 15.1 Summary of soil loss by water erosion measured in different studies using runoff plots or at small watersheds under natural rainfall

Experiment	Soil management	Soil loss (t ha^{-1} $year^{-1}$)	Duration (years)	Plot length (m)	Slope (%)	Plot size (m^2)
Conchuela I	NLH	6.9	5	12	13.4	72
	LC	2.9	5	12	13.4	72
	CV	0.8	5	12	13.4	72
Benacazon	LC	29.6	6	60	11	480
	CV	58.9	3	24	13.4	288
Conchuela II	LC	5.6	2	24	13.4	288
	CV	5.6	2	24	13.4	288
Lanjarón I	NLH	25.6	2	24	30	480
	LC	5.7	2	24	30	480
	DC	2.1	2	24	30	480
Lanjarón II	NT	25	2	22.5	30.5	480
	CT	8.5	2	22.5	30.5	480
	DC	6.5	2	22.5	30.5	480

Experiment	Soil management	Soil loss (t ha^{-1} $year^{-1}$)	Duration (years)	Watershed length (m)	Average slope (%)	Watershed area (m^2)
Conchuela	CT and CC	16.1	5	270	9	80,000
Setenil	NT and CT	9.3	6	340	9.5	67,000
Puente Genil	DC	1.8	5	150	15	61,000

NLH is bare soil with herbicide and no tillage; LC, bare soil with tillage; and CV, temporary cover crop Adapted from Gómez et al. (2014)

These experimental studies, combined with sampling from commercial farms with different soil management practices, have provided an overview of the impact of these different management methods on soil properties. Figure 15.4 shows how cover crop provides a significant improvement in various soil quality indicators in the surface soil horizon, approximately 0–10 cm. These differences between management practices gradually disappear as with depth. In many cases, they disappear or are only minor below a depth of approximately 20–30 cm.

To understand the how relevant can be this positive impact of soil management on mitigating olive orchard soil degradation, it is necessary to understand that it depends on the area covered, the biomass produced, and the duration of the cover crop in the field. Although Fig. 15.4 indicates that there is appreciable potential for improving soil quality and reversing degradation, the need to control canopy transpiration to avoid excessive competition with the tree implies the existence of an operational maximum for variables that depend greatly on this biomass production, such as carbon sequestration (Vicente-Vicente et al. 2016).

Given the extent and diversity of olive groves in Spain, the best way to assess the state of soil erosion and degradation is through simulation models. The most studied aspect is water erosion at the hillside scale using the Universal Soil Equation or its updated version, the Revised Universal Soil Equation. An example of this work is that of Gómez and Giráldez (2009), who conducted an analysis for Andalusia considering different scenarios (Fig. 15.5).

Given the soil and climate conditions and the experimental results available (Table 15.1), the results depicted in Fig. 15.5 seem plausible and illustrate two realities. First, the widespread use of temporary vegetation cover, initially favored and later imposed on sloping areas by the Common Agricultural Policy (CAP), has had a positive effect

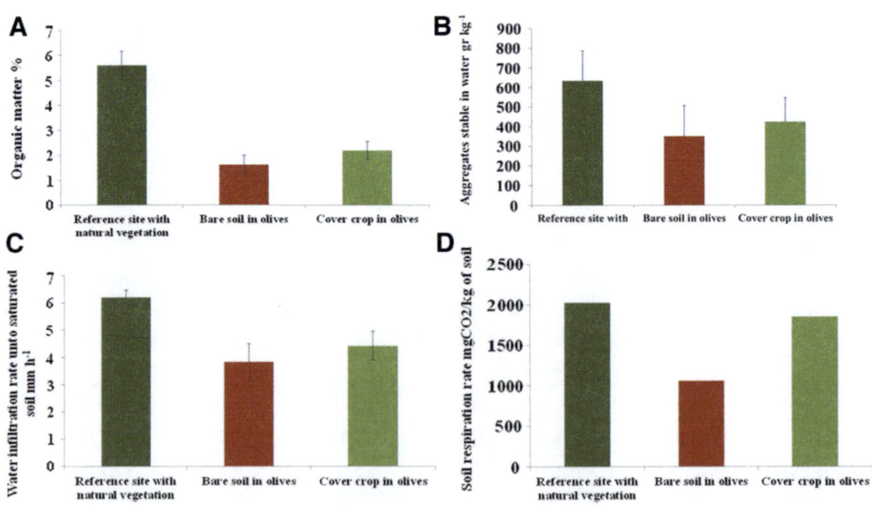

Fig. 15.4 Summary of some soil properties related to soil quality in relation to soil management system. Adapted from Gómez et al. (2014)

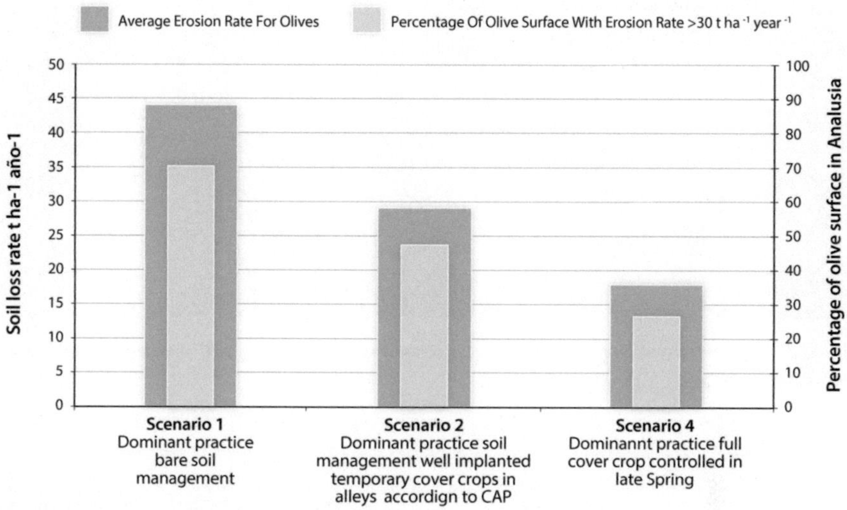

Fig. 15.5 Summary of water erosion estimates in olive orchards in Andalusia under different soil management scenarios. Adapted from Gómez and Giráldez (2009)

on reducing water erosion rates in olive orchards in Andalusia, something that could be generalized to the rest of Spain. Second, despite this, olive grove areas with erosion rates above the tolerable threshold remain, even using an undemanding threshold that is common in other RUSLE/USLE-based analysis. This later result applies areas where olives are still grown on steep slopes; see Fig. 15.2. This analysis, and any other based on the extrapolation of olive grove areas using vegetation cover, must be interpreted with caution for several reasons. The first of them is that the effect on erosion control of the vegetation cover depends on its correct implementation so that it provides good soil cover, and various works in olive orchards (e.g., Vicente-Vicente et al. 2016) and other woody crops (such as vineyards, e.g., Guzman et al. 2019) indicate a very high variability between farms, see Fig. 15.6.

A second reason, is that these large-scale analyses hide very different realities at the regional level, which can vary greatly depending on local conditions. An example is the analysis carried out for the Estepa PDO, with approximately 40,000 ha of olive groves, using the same methodology. In this study, using a similar analysis, it was estimated that 89% of the total olive grove area presented soil losses due to water erosion at the hillslope scale of less than 10 t ha^{-1} $year^{-1}$ (Gómez et al. 2023a). The third caveat is that other erosive processes, such as gully erosion, also intense in some olive grove areas (Fig. 15.7), are not included in these model analyses. Nevertheless, they provide an overview of the intensity of water erosion problems and their variability across the region.

The big question is, if there is a serious erosion problem in olive groves, why hasn't it been addressed yet? The author of this chapter doesn't have the final answer, but

Fig. 15.6 View of two olive groves in winter, both with temporary vegetation cover. Photos by José Alfonso Gómez

Fig. 15.7 Image of a gully in an olive grove. Photo by José Alfonso Gómez

he can contribute with some of the pieces of this puzzle that remain to be completed. One of them is that management with temporary cover crops, although it presents many environmental (and sometimes agronomic or management) advantages, requires proper implementation to avoid the risk of crop loss due to competition with the tree. An analysis of the available experimental data indicates that this crop loss has been measured in a significant number of experiments, around 20% (see Fig. 15.8). This yield loss is in line with what has been detected with water balance models, which worsens compared to the system with bare soil when this cover control is not carried out at the appropriate time (e.g., Abazi et al. 2013; Lorite et al. 2025).

A third piece of the puzzle is that in many situations, olive grove productivity does not show a dramatic decline, but rather, through a combination of agronomic improvements and soil resilience, production levels have been maintained or increased. An example of this was documented in the historical erosion work of Vanwalleghem et al. (2011). In the three olive orchards studied, while approximately 45 cm of soil was lost over an average period of about 222 years, the olive yield increased from 400 kg ha^{-1} $year^{-1}$ to 2500 kg ha^{-1} $year^{-1}$, approximately. This does not mean that the soil's productive capacity has not been reduced (which has occurred but is hidden), nor that this degradation can be maintained indefinitely. But it does indicate that, to date, this degradation has not manifested itself alarmingly in production losses, with a few exceptions (see Fig. 15.9).

In this context, a soil degradation problem whose importance and magnitude are recognized by farmers (e.g., Gómez et al. 2021) but which is not pressing in a context of difficult economic viability for many farms, becomes a problem whose approach is postponed year after year. This is compounded by the technical difficulty of defining appropriate management at the farm scale and the absence of an comprehensive advisory system for farmers (especially small and medium-sized ones).

Successfully addressing the fight against desertification in olive orchards will require a combination of socioeconomic and technical measures focused on the farm level, where the general understanding generated over the decades, while useful, is insufficient to implement operational solutions tailored to specific edaphoclimatic and socioeconomic conditions.

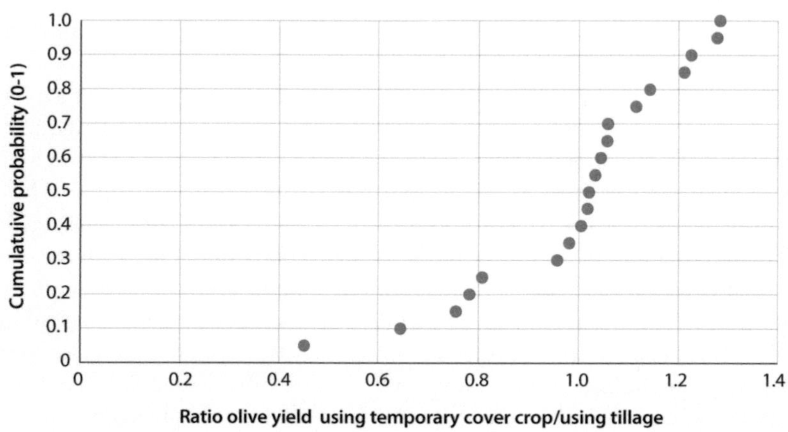

Fig. 15.8 Cumulative distribution probability of the ratio of the average olive yield in olives orchards using cover crops divided by the average olive yield in tilled olive orchards. Adapted from Gómez et al. (2023b)

Fig. 15.9 Extremely degraded olive grove in the province of Cordoba. Photo by José Alfonso Gómez

Pressure on Water Resources

Irrigation in olive cultivation has experienced a large increase since the early 1980s (see Fig. 15.10), with approximately 600,000 ha to date. It is worth noting that these orchards are irrigated under deficit irrigation, which is irrigating below the full requirements of the crops. This results with an average water use for irrigation of 1500 m^3 $year^{-1}$, or 150 mm $year^{-1}$, (MAPA 2024), which is applied using localized (drip) irrigation systems during the dry season. Its impact on yield is also depicted in Fig. 15.10, which indicates how the gap between olive orchards with and without access to irrigation has been widening in last years.

Olive tree yield is very sensitive to any increases, even small, in water availability (Moriana et al. 2003) which makes it an attractive option for deficit irrigation. This increase in yield, and in potential income for farmers (e.g., Mesa-Jurado et al. 2010) has been behind this expansion. Although part of this expansion of irrigation has been made using water previously allocated to other crops, in overall terms has resulted in an increased pressure in water resources, which is particularly acute in areas which were already under pressure. This has resulted in some areas in an exacerbation of groundwater overexploitation with aquifer depletion and ecological damage to surface wetlands. This impact can be amplified by climate change, which is predicted to reduce water availability for olives (e.g., Abazi et al. 2013; Lorite et al. 2025). This pressure on water resources needs to be accounted for future actions to fight against desertification in olive-growing areas. Until now, most of the efforts have been oriented towards increasing water efficiency, which is already very high (Arbizu-Milagro et al. 2023). But it should be integrated with other actions like: stricter enforcement and governance;

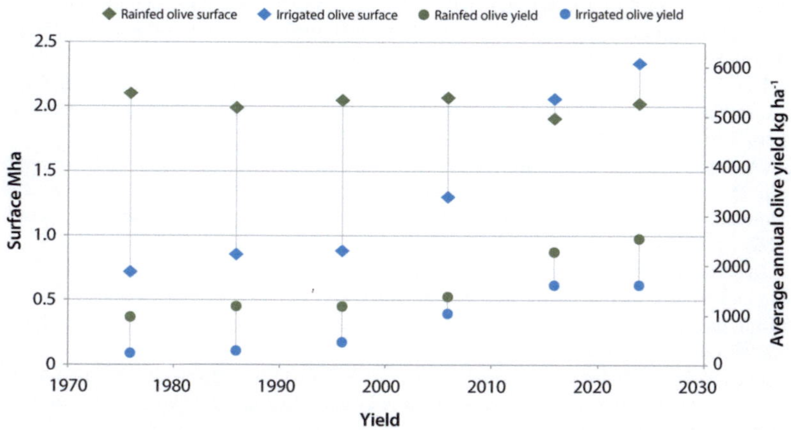

Fig. 15.10 Evolution of surface and average yield of rainfed and irrigated olives in Spain since 1976. Source: Own elaboration from MAPA (2025)

supporting traditional rainfed production valuing and supporting high-quality, sustainable rainfed olive oil production through premiums and market differentiation, compensating farmers to revert irrigated land to dryland or less water-intensive uses, or developing drought-resistant olive varieties (a long-term solution) and integrating climate projections rigorously into all these measures.

Acknowledgments Elaboration of this chapter has been financially supported by the project RELAND (PID2023-146177OB-C21 funded by MICIU/AEI/10.13039/501100011033) and by "ERDF A way of making Europe."

References

Abazi U, Lorite IJ, Cárceles B, Martínez Raya A, Durán VH, Francia JR, Gómez JA (2013) WABOL: a conceptual water balance model for analyzing rainfall water use in olive orchards under different soil and cover crop management strategies. Comput Electron Agric 91:35–48. https://doi.org/10.1016/j.compag.2012.11.010

Álvarez S, Soriano MA, Landa BB, Gómez JA (2007) Soil properties in organic olive orchards as compared to natural areas in a mountainous landscape in Southern Spain. Soil Use Manag 23:404–416. https://doi.org/10.1111/j.1475-2743.2007.00104.x

Arbizu-Milagro J, Castillo-Ruiz FJ, Tascón A, Peña JM (2023) Effects of regulated, precision and continuous deficit irrigation on the growth and productivity of a young super high-density olive orchard. Agric Water Manag 286:108393. https://doi.org/10.1016/j.agwat.2023.108393

Beaufoy G (2001) EU policies for olive farming. Unsustainable on all Counts; Bird Life International-WWF, 20 pp, https://www.wwf.eu/?8818/EU-Policies-for-olive-farming-unsustainable-on-all-counts

Bennett HH (1960) Soil erosion in Spain. Geographical Rev 50:59–72

De la Fuente L, Navas-Cortés JA, Landa BB (2024) Ten challenges to understanding and managing the insect-transmitted, xylem-limited bacterial pathogen Xylella fastidiosa. Phytopathology 114:869–884. https://doi.org/10.1094/PHYTO-12-23-0476-KC

Gómez JA (ed) (2009) Sostenibilidad de la producción de olivar en Andalucía, Sevilla. Consejería de Agricultura y Pesca, Junta de Andalucía, Sevilla, p 317, https://www.juntadeandalucia.es/export/drupaljda/1337160963Sostenibilidad_agraria.pdf

Gómez JA Giráldez JV (2009) Erosión y degradación de suelos. In: Gomez JA (ed) Sostenibilidad de la Producción de Olivar en Andalucía, Consejería de Agricultura y Pesca. Junta de Andalucía, Sevilla, p 42, https://www.juntadeandalucia.es/export/drupaljda/1337160963Sostenibilidad_agraria.pdf

Gómez JA, Infante-Amate J, González de Molina M, Vanwalleghem T, Taguas EV, Lorite I (2014) Olive cultivation, its impact on soil erosion and its progression into yield impacts in Southern Spain in the past as a key to a future of increasing climate uncertainty. Agriculture 4:170–198. https://doi.org/10.3390/agriculture4020170

Gómez JA, Sánchez Montero A, Guzmán G, Soriano MA (2021) In-depth analysis of soil management and farmers' perceptions of related risks in two olive grove areas in southern Spain. Int Soil Water Conserv Res 9:461–473. https://doi.org/10.1016/j.iswcr.2021.01.003

Gómez JA, Sánchez A, Soriano MA, Guzmán MG (2023a) Evaluation of erosion risk with stakeholders using RUSLE methodology and publicly available information in a large olive producing area in Southern Spain. Paper presented at the international symposium Soil Erosion Research under a Changing Climate Symposium, ASABE, Aguadilla, Puerto Rico

Gómez JA, Guzman MG, Lorite I, Soriano MA (2023b) Cubiertas vegetales, balance de agua y efecto en cosecha con respecto a laboreo: Una mirada a información generada durante décadas. Paper presented at the international symposium XXI Simposio Científico-Técnico Expoliva 2023, Jaén, España

Guzmán JR (2004) El Palimpsesto cultivado. Consejería de Agricultura y Pesca, Junta de Andalucía, Sevilla, p 439, https://www.juntadeandalucia.es/export/drupaljda/1337165052El_Palimpsesto_cultivado.pdf

Guzmán JR, Hernández P, Gómez JA, Lora A (2020) Olivares de España recorrido por la biografía del olivar, su memoria y sus paisajes, Almuzara, Córdoba, 992 p

Guzmán G, Cabezas JM, Sánchez-Cuesta R, Lora Á, Bauer T, Strauss P, Winter S, Zaller JG, Gómez JA (2019) A field evaluation of the impact of temporary cover crops on soil properties and vegetation communities in southern Spain vineyards. Agric Ecosyst Env 272:135–145. https://doi.org/10.1016/j.agee.2018.11.010

Hayas A, Vega R, Gómez JA (2025) Analyzing gully evolution in a rolling landscape over seven decades in SW Europe. Lessons for the future. Geoderma https://doi.org/10.2139/ssrn.5115341

Hernández MP (2011) Clasificación territorial del olivar español. Dissertation, University of Cordoba

Infante-Amate J, González de Molina M, Vanwalleghem T, Soto Fernández D, Gómez JA (2014) Reconciling Boserup with Malthus: agrarian change and soil degradation in olive orchards in Spain (1750 – 2000). In: Fischer-Kowalski M, Reenberg A, Schaffartzik A, Mayer A (eds) Ester Boserup´s Legacy on Sustainability. Orientations for contemporary, human-environmental interactions, Vol 4, Springer, p 18, https://doi.org/10.1007/978-94-017-8678-2

Lorite I, Cabezas JM, Soriano MA, Alza J, Santos C, Gómez JA (2025) Analysing the role of soil management in the sustainability of Mediterranean olive groves under current and future weather conditions (in revision process in Spanish Journal of Agricultural Research)

MAPA (2024) Encuesta sobre superficies y rendimientos cultivos (ESRYCE). Encuesta de Marco de Áreas de España. Ministerio de Agricultura, Pesca y Alimentación https://www.mapa.gob.es/es/estadistica/temas/estadisticas-agrarias/agricultura/esyrce/ (accessed 30 May 2025)

MAPA (2025) Estadisticas Agrarias. Ministerio de Agricultura, Pesca y Alimentación. https://www.mapa.gob.es/es/estadistica/temas/estadisticas-agrarias/ (accessed 30 May 2025)

Mesa-Jurado MA, Berbel J, Orgaz F (2010) Estimating marginal value of water for irrigated olive grove with the production function method. Span J Agric Res 8:197–206. https://doi.org/10.5424/sjar/201008S2-1362

Montes-Osuna N, Mercado-Blanco J (2020) Verticillium wilt of olive and its control: what did we learn during the last decade. Plants 9:735. https://doi.org/10.3390/plants9060735
Moriana A, Orgaz F, Pastor M, Fereres E (2003) Yield responses of a mature olive orchard to water deficits. J Am Soc Hortic Sci 128:425–431. https://doi.org/10.21273/JASHS.128.3.425
Rodrigáñez C (1908) La Tierra que se pierde. El Imparcial, Madrid, Spain
Ruiz de Castroviejo J (1969) Ensayos realizados en la provincia de Córdoba sobre explotación de olivares en asociación con trébol subterráneo. Agricultura 443:135–139
Scheidel A, Krausmann F (2011) Diet, trade and land use: a socio-ecological analysis of the transformation of the olive oil system. Land Use Policy 28:47–56. https://doi.org/10.1016/j.landusepol.2010.04.008
Vanwalleghem T, Infante J, González de Molina M, Soto D, Gómez JA (2011) Quantifying the effect of historical soil management on soil erosion rates in Mediterranean olive orchards. Agric Ecosyst Env 142:341–351. https://doi.org/10.1016/j.agee.2011.06.003
Vicente-Vicente JL, García-Ruiz R, Francaviglia R, Aguilera E, Smith P (2016) Soil carbon sequestration rates under Mediterranean woody crops using recommended management practices: a meta-analysis. Agric Ecosyst Env 235:204–214. https://doi.org/10.1016/j.agee.2016.10.024

Chapter 16
Hiding Desertification with Technology: Irrigation and Water Imbalance in Southern Spain

Gonzalo G. Barberá and Víctor Castillo

Abstract In this chapter, we briefly discuss the effects of the expansion of new, intensive irrigated agriculture on the desertification processes in the Mediterranean. First, we examine how the intensification in the new irrigated lands leads to desertification, while at the same time, the process itself "hides" its own consequences. We then present how the in situ degradation processes work. However, while locally the degradation is hidden by the intensification, we show how the ongoing desertification processes are indirectly reflected in larger-scale water dynamics, both on surface hydrology and on groundwater, sometimes with paradoxical effects. This mechanism is illustrated with three paradigmatic case studies from South Spain, Doñana National Park, Tablas de Daimiel National Park and Mar Menor Lagoon. Finally, we discuss why the desertification debate is easily distorted by the hidden effects behind "greening fields" and how land degradation induced by new irrigated lands can be addressed.

Keywords Drip irrigation · South Spain · Soil degradation · Hydrological cycle · Landscape simplification · Agricultural intensification

G. G. Barberá (✉) · V. Castillo
Department of Soil and Water Conservation, Centro de Edafología y Biología Aplicada del Segura (CEBAS), CSIC, Murcia, Spain
e-mail: gbarbera@cebas.csic.es

V. Castillo
e-mail: victor@cebas.csic.es

J. Martínez-Valderrama and J. Olcina Cantos (eds.), *The Labyrinth of Desertification*,
https://doi.org/10.1007/978-3-032-09648-7_16

The Classic Paradigm(s) of Desertification and the Agricultural Intensification as a Trompe L'oeil

Desertification is not always a clear-cut concept or even a monosemic one. The basic definition should be sourced from the Article 1 of the UNCCD (*see* Chap. 1). This definition is very general, but the underlying basis to it is the transition from non-desert to desert-like conditions (Sterk et al. 2016), a reduction of land productivity, in areas where the evapotranspiration is higher than precipitation. Note that the effect and the environmental setting are clearer than cause as it includes "climatic variations," but also generic "human activities."

The conventional paradigm of desertification in the Mediterranean focuses on land degradation triggered by soil deterioration due to agricultural mismanagement, overgrazing, or deforestation. In this view, soil quality and protection—largely dependent on organic carbon content—are central. However, in water-limited ecosystems with naturally low productivity, soil organic carbon is typically scarce. Any disturbance that reduces vegetation cover (e.g., converting natural land to agriculture) or depletes soil organic matter through repeated plowing leads to degraded soils with reduced infiltration, water retention, and nutrient availability. This sets off a self-reinforcing cycle of degradation that can ultimately result in desertification. However, this traditional model is increasingly being challenged by new forms of intensive agriculture that rely heavily on external inputs, altering the dynamics of degradation and masking its early symptoms.

Agriculture, even in its most traditional forms, is a technology that alters environmental conditions to enhance crop yields, relying on external inputs such as labor, energy and fertilizers. Intensification increases dramatically these inputs, particularly in arid regions, where irrigation becomes essential. Historically, irrigation relied on gravity and flat terrain, but modern technologies now allow irrigation on steeper slopes and more complex landscapes. In southeastern Spain, the expansion of irrigated lands over the past 50 years has led to aggressive transformations, replacing both natural ecosystems and rainfed agriculture—a process that has accelerated with climate change and its consequences, particularly in the form of prolonged droughts. Heavy machinery flattens terrain (see Fig. 16.1), removes boundaries, and destroys protective soil features like petrocalcic horizons. The resulting plots prioritize mechanization and productivity, often at the expense of soil health. Soil conservation practices are frequently abandoned, further increasing erosion risks. Drip irrigation dominates these new systems, and let to apply to steep, mechanically altered slopes, and, therefore, extended soil degradation, making intensification a driver of desertification in vulnerable regions.

Paradoxically, the effects of these desertification-related processes remain hidden, as there is no clear or immediate decline in land productivity. In fact, the intensive use of groundwater—on which much of the so-called economic miracle of this new agriculture relies—has been referred to as the "Silent Revolution" (Llamas and Martínez-Santos 2005), precisely because the resource itself is invisible. Drip irrigation is particularly suitable to supply the crops with fertilizers dissolved in irrigation water and, therefore, plants receive water and nutrients in adequate quantities irrespective of the soil quality status. In fact, in these new irrigated lands, soils are of very poor quality—effectively desertified—but the visible impacts of desertification are masked.

Fig. 16.1 **A** Reforestation with pine trees, following the contours of the small reliefs, and irrigation olive orchards in the study area (2004). **B** The same area in 2019, where afforestations have been dismantled and the territory has been levelled to facilitate the planting super intensive irrigated olive groves. Map data: Google, DigitalGlobe. Adapted from Martínez-Valderrama et al. (2020)

Thanks to drip irrigation, water and nutrients are supplied directly to the crops, making productivity largely independent of soil quality, rainfall infiltration, or natural nutrient availability. Some indirect evidence of this mechanism is the fact that from 1962 to 2020, 52% of the increase of agricultural production in Spain is attributable to the yield irrigation effect while only 22% is attributable to the effect of irrigated area (Serrano et al. 2024). This decoupling between soil quality and crop productivity may be compared to the *trumpe d'oleil*, a pictorial technique that attempts to deceive the eye, aiming to the substitution of reality by the created optic effect.

The Expansion of the Irrigated Agriculture in South Spain and Its Impacts on Local Soils

Over the last century, irrigated agriculture has expanded globally, especially in regions where evapotranspiration exceeds precipitation—namely, drylands, the only areas where desertification can occur. In Spain, irrigated land occupied a stable area of 1.3–1.4 Mha on the first decades of the twentieth century (1900–1940), but after the adoption of diverse public policies, the irrigated area reached 2.2 Mha by 1970 and 3.7 Mha by 2000, with some stabilization afterwards (García-Mollá et al. 2019). However, between 2004 and 2023, a further increase of 0.4–0.5 Mha has been reported (Arbonès et al. 2024). Not all this increase implies the heavy land transformation like described in the previous section.

García-Ruiz et al. (2013) noted that the risk of desertification from an erosion perspective was low in irrigated lands, as these were typically located on plots with gentle slopes. However, this constraint was overcome with the introduction of drip irrigation, which enabled irrigation even on steeper terrain. Also, the authors identified "plot concentration" as a problem because of the removal of internal barriers in the fields, as was introduced in the previous section.

Surprisingly, there is limited quantitative information about soil erosion and degradation in new irrigated lands under intensive agriculture in the Mediterranean. By contrast, there is a wealth of information for traditional dominant Mediterranean crops, such as olive trees, vineyards or cereals, as well as for abandoned agricultural lands. Furthermore, the term "intensive agriculture" is often applied to crops with very different actual intensity. For instance, non-irrigated, non-fertilized but tilled olive tree plantation is labeled "intensive agriculture" (Lucas-Borja et al. 2019).

To avoid confusion among farming systems with differing impacts, we refer to "intensive agriculture" as systems characterized by technical irrigation, the use of chemical fertilizers, and typically high levels of herbicide and pesticide application. It is important to note that organic agriculture can also adopt intensive modes of production that can affect soil health in ways similar to conventional agriculture (Van Rijssel et al. 2025) although conventional herbicides, pesticides, or synthetic fertilizers are not used. As mentioned in the previous section, this kind of intensive agriculture in the Mediterranean is also associated to frequent machinery use and, in a considerable number of cases, with drastic landscape reshaping to accommodate planting and irrigation infrastructures.

The impact of these new irrigated lands on the soil is significant and fast. In tree crops such as citrus (see Fig. 16.2) or stone fruit orchards, machinery moves on

Fig. 16.2 New citric plantation over highly erodible marls in semiarid SE Spain. Sloping terrain with minimum organic matter content and without any soil conservation measure. Photo by Pedro Ruiz

alleys between tree rows, increasing soil compaction. Cerdà et al. (2021b) found an increase on bulk density from 1.05 to 1.33 g cm^{-3} over 13 years in a citrus orchard. In a peach tree orchard, the bulk density under the trees was 1.05 g cm^{-3} but increased to 1.23 g cm^{-3} where the machinery circulated (Cerdà et al. 2021a).

An increase in bulk density may also be reinforced by the application of herbicides on the alleys to reduce competition between weeds and trees, resulting in bare surfaces with no organic carbon supply to the soil. A comparison between conventionally managed orange orchards where herbicide was used and organic managed shows four times higher soil organic matter and total nitrogen in the latter, as well as substantial differences in available P and soil aggregate stability (Morugán-Coronado et al. 2015). The abandonment of citrus orchards drastically reduces the soil erosion from 607 to 7 g m^{-2} (Cerdà et al. 2019), clearly indicating the soil degradation effects of intensive agriculture.

Hidden Desertification at the Field Scale Become Manifest at Landscape and Catchment Scales Through Changes in Water Dynamics

In Spain, the development of forest policy since the late nineteenth century was closely linked to the control of hydrological regimes altered by deforestation, overgrazing and cultivation of marginal lands (Pemán García 2024). The classic example of reforestation par excellence in Spain is Sierra Espuña, in SE Spain (1895–1924). This reforestation was prompted by the catastrophic flood of *Santa Teresa* in 1879. It was recognized that deforestation in the Segura basin was largely responsible for the magnitude of the flood (López Azorín 2021). The subsequent development of reforestation policy in Spain for the next century prioritized "forest hydrological restoration projects" aimed to protect headwater reservoir catchments from torrential rainfall. These projects focused on hillslope reforestation and building check dams in stream channels.

Groundwater overexploitation is another mechanism of alteration of the hydrologic cycle that has been included in the human actions inducing desertification. The association is more recent as overexploitation of aquifers has been only possible since twentieth century. Martínez-Fernández and Esteve (2005) already remarked that desertification processes in SE Spain were mainly associated to the development of new irrigated lands supported by groundwater extraction (see Chap. 14).

Thus, the link between desertification at local or landscape scales and altered hydrologic and hydrogeological dynamics at larger scales is not a new idea at all but, as we will see later, new twisted narratives about desertification are flipping the argument, suggesting that further forcing the hydrological cycle is the best way to mitigate desertification.

At present, three important environmental crises are unfolding in Spain, all related to these altered hydrological cycles. Each crisis affects to the one among most emblematic natural areas of the country (Fig. 16.3).

Tablas de Daimiel is a classic example of a groundwater-dependent wetland. Since the 1970s, the expansion of irrigated agriculture has severely depleted the underlying

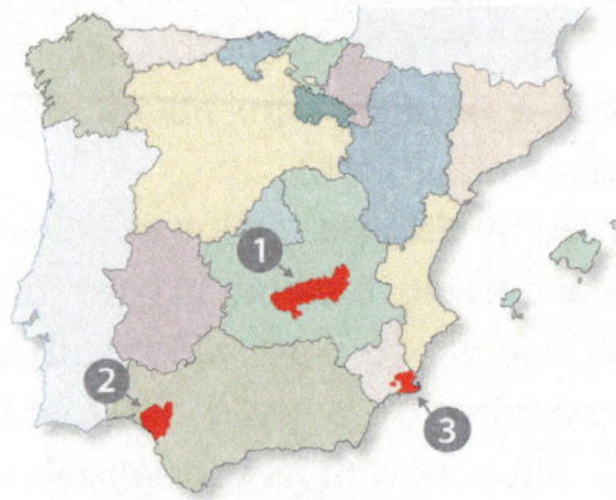

	Extraction volume (hm^3)	Irrigated area (ha)	Basin surface (km^2)
1 **Tablas de Daimiel National Park** (Mancha Occidental I, Mancha Occidental II and Rus-Valdelobos aquifers)	**92.39**	**51,465**	**6,117**
2 **Doñana National Park** (Doñana and Aljarafe aquifers)	**21.13**	**4,729**	**2,845**
3 **Mar Menor lagoon** (Campo de Cartagena aquifer)	**36.00**	**9,100**	**1,316**

Fig. 16.3 Location of three emblematic wetland areas in Spain: Tablas de Daimiel, Doñana, and Mar Menor. Tablas de Daimiel and Doñana are designated National Parks—the highest level of environmental protection in Spain—while Mar Menor holds multiple, though more fragmented, protection statuses. Despite all three being wetland systems that have experienced significant expansion of irrigated agriculture within their catchments and over associated aquifers. For Tablas de Daimiel and Doñana, the figures show the irrigated areas and the estimated volume of extracted water in zones located outside the legal boundaries of the Agricultural Demand Units (for 2020 and 2019 in the case of Tablas de Daimiel), as well as the surface area of aquifers that are being illegally exploited. The data for the Mar Menor corresponds to illegal irrigation systems closed as of June 2025, supplied not exclusively with underground resources, with an estimated allocation of 4000 m^3 ha^{-1}. Sources: WWF (2021), https://www.miteco.gob.es/es/prensa/ultimas-noticias/2025/junio/la-labor-de-control-del-regadio-ilegal-eleva-a-9-100-lashectare.html

aquifer, leading to the desiccation of the wetland. Partial recovery has been observed following the implementation of restrictions on groundwater extraction (Castaño et al. 2018). Doñana encompasses a diverse array of wetlands that depend on both rainfall and surface runoff from the surrounding catchment, as well as groundwater discharge from an aquifer that extends beyond the boundaries of the national park. Some wetland types rely more heavily on surface water, while others depend primarily on groundwater (Olías et al. 2025). Since the 1970s, water has been extracted from the aquifer, and in recent decades, the expansion of strawberry and other berry crops has led to a significant increase in water withdrawals. This has resulted in declining piezometric levels, negatively impacting groundwater-dependent wetlands, river inflows into the park, and even surface evapotranspiration (Olías et al. 2025).

In both cases, groundwater overexploitation—primarily to support new irrigated lands—has severely disrupted the hydrological cycle, with profound impacts on natural ecosystems. While this may seem distinct from the conventional desertification paradigm, the two are intrinsically linked. Water is the key limiting factor in both agricultural and natural ecosystems in drylands, where desertification can occur. Soil degradation through erosion and loss of organic carbon impairs multiple functions, most critically its ability to retain and supply water to crops. This leads to increased runoff and altered micro-scale hydrology. The expansion of irrigated lands aims to enhance crop yields by accessing water at larger spatial scales. However, such manipulation often disrupts the hydrological cycle far beyond the fields themselves, with consequences at the regional scale.

The Mar Menor lagoon case represents, paradoxically, the opposite process. It was originally a semiarid area with no permanent water courses and a Quaternary aquifer in contact with the lagoon, whose natural discharge was low and very poor in nutrients. Since the 1980s, irrigated agriculture has expanded 10-fold supported by a water transfer from Tagus River direct and indirectly, as there is an allocation for agriculture and another for urban uses, but the urban water is regenerated and reused for agriculture.

More recently, seawater desalination has been incorporated into the system. As a result, the hydrological alterations driven by the expansion of irrigated lands in the Mar Menor region (and elsewhere in Southeastern Spain) now occur at a much broader scale than in Tablas de Daimiel or Doñana. Instead of a local reduction in water discharge to wetlands, as seen in those cases, the Mar Menor area has experienced a significant increase—yet one that negatively impacts natural ecosystems. Both surface and groundwater discharges have risen. Although historical data are limited, clear evidence includes the emergence of permanent watercourses caused by rising piezometric levels. This groundwater is heavily contaminated with nitrates due to the prolonged and intensive use of fertilizers in high-input agricultural systems.

This disruption has played a key role in the recurring eutrophic crises since 2016 (Álvarez-Rogel et al. 2020). Additionally, the persistent and elevated submarine groundwater discharge has triggered an unusually long-lasting whiting event in parts of the lagoon (Oosterbaan et al. 2025). One of the most striking aspects of the Mar Menor case is the complexity of impacts directly linked to the expansion of new irrigated lands. While groundwater discharge is a major concern, surface hydrology and water quality during intense rainfall events have also been significantly affected. These issues stem largely from hidden desertification: soils are severely degraded, with poor structure and low organic content, and landscape simplification has increased hydrological connectivity, resulting in more intense and rapid runoff responses.

Salinity levels in the lagoon have become increasingly variable in recent decades, likely due to enhanced runoff during storm events (Jiménez Casero 2024). Floods can carry extremely high nutrient loads, driven by the erosion of sediments and the release of nutrients accumulated in soils after decades of intensive farming (Álvarez-Rogel et al. 2020). Ultimately, while desertification may remain invisible at the field scale, the broad disruption of hydrological processes reveals the severity of the degradation at the landscape level.

Box 16.1: Hidden Desertification and New Narratives About How to Combat It

The emergence of new irrigated lands and the "hidden" nature of some desertification processes have weakened the perceived link between land degradation and desertification—at least for a significant part of the public. Much of this disconnect stems from the emotional weight carried by the term "desertification." Its evocative power, drawing on associations with "desert" and aridity, likely explains why the issue has received such global attention, including a dedicated UN convention.

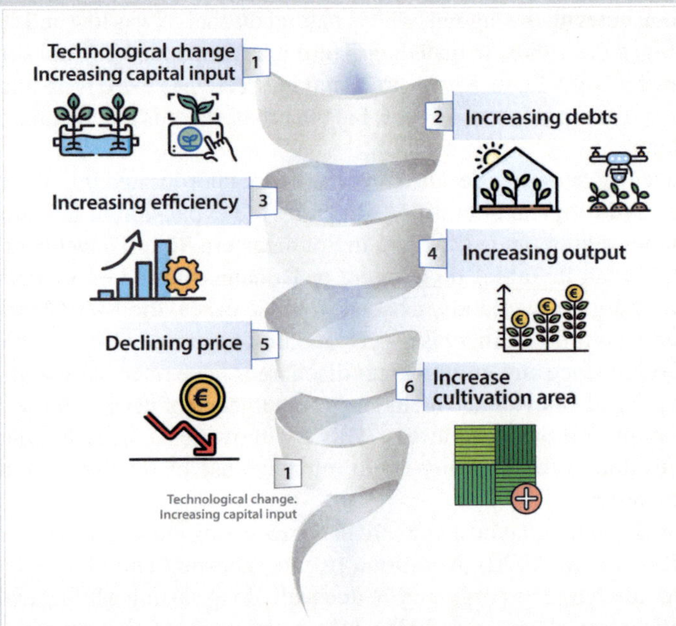

The Treadmill of Production operates through a repeating six-step cycle: (1) Increasing capital inputs; (2) Rising debt levels; (3) Improved yields/efficiency; (4) Increased output; (5) Falling product prices; and (6) Expansion of cultivated land. While these investments aim to maximize short-term profitability, they generate numerous negative externalities, including dependence on external inputs, precarious labor conditions, food waste due to overproduction, inequitable wealth distribution, and significant environmental degradation Source: Adapted from Martínez-Valderrama et al. (2023)

Now consider a scenario where actual degradation of soil and landscape is masked by intensive inputs of water and technology. In such contexts, lush green crops flourish amid degraded brown surroundings—a modern oasis. What better symbol of resisting desertification than an oasis? This narrative suggests that desertification can be defeated by continuously expanding irrigation. But this feeds a cycle of environmental degradation and social precarity, as technological investments grow and only the most financially resilient actors can remain competitive.

This is not merely theoretical. In southeastern Spain, this belief is widespread—reflected in media, politics, water user associations, and even academia. Irrigation enjoys overwhelming public support, associated almost exclusively with benefits and progress.

Narratives like these—simple, emotionally resonant, and easily amplified online—pose a significant challenge. If combating desertification is the goal, the academic and institutional communities must critically engage with and challenge these dominant narratives.

What to Do to Combat Desertification Brought About by New Intensively Irrigated Lands? Barriers and Opportunities

Agrarian production from new irrigated lands is not going to disappear. A thorough discussion of all potential impacts—both environmental and social, whether positive or negative—as well as the role of global trade, consumer preferences and behavior, and other externalities, is beyond the scope of this chapter. The necessity and relevance of intensive agriculture is another example of the dilemmas that society faces when trying to expand economic activities while reducing natural resource use and environmental and social impacts. In this chapter, we simply take as a starting point that, to a greater or lesser extent, this type of agrarian production will continue to exist.

Even if this assumption is accepted, it is clear that an *ad infinitum* expansion on irrigated lands is not feasible. Water resources in South Spain, the Mediterranean, and most areas prone to desertification are limited. Adding non-conventional water resources like sea desalinated water is complex due to its high cost. In agriculture, desalinated water is usually subsidized raising concern about long-term availability.

An agreement about the limits of irrigation expansion is urgently needed, but unfortunately seems unlikely. In Spain, at least, several factors hinder such consensus: the widespread cultural belief that irrigation is a necessity, the large profitability gap between rainfed and irrigated agriculture, the European-wide increase in market demand for fruits and vegetables and interregional competition for water resources. From this perspective, the adjustment seems more likely to come through *force majeure*, that is, the exhaustion of water resources, leading to a widespread and abrupt unhiding of desertification across the country.

The paragraph above describes the macro scale of the problem. At the micro scale, the perspectives are much better. It is possible to act at farm scale in three lines: reducing inputs adopting ecological intensification (Kleijn et al. 2019), improving organic matter content, implementing soil conservation measures adapted to the reality of new irrigated agriculture. These actions often require more ingenuity than financial investments, and their success largely depends on a combination of cultural change in the perspective of managers and the pressure from environmentally concerned consumers through the supply chain. In our present experience, this change, albeit still little, is already underway and gaining momentum.

References

Álvarez-Rogel J, Barberá GG, Maxwell B et al (2020) The case of Mar Menor eutrophication: state of the art and description of tested nature-based solutions. Ecol Eng 158:106086. https://doi.org/10.1016/j.ecoleng.2020.106086

Arbonès G, De Stefano L, Garrido A (2024) Agricultural land use change and intensification in Spain: drivers and trends (2004–2023). Available at SSRN: https://ssrn.com/abstract=4963512 or https://doi.org/10.2139/ssrn.4963512

Castaño S, De La Losa A, Martínez-Santos P et al (2018) Long-term effects of aquifer overdraft and recovery on groundwater quality in a Ramsar wetland: Las Tablas de Daimiel National Park, Spain. Hydrol Process 32:2863–2873. https://doi.org/10.1002/hyp.13225

Cerdà A, Ackermann O, Terol E, Rodrigo-Comino J (2019) Impact of farmland abandonment on water resources and soil conservation in citrus plantations in Eastern Spain. Water 11:824. https://doi.org/10.3390/w11040824

Cerdà A, Daliakopoulos IN, Terol E et al (2021a) Long-term monitoring of soil bulk density and erosion rates in two Prunus persica (L) plantations under flood irrigation and glyphosate herbicide treatment in La Ribera district, Spain. J Env Manage 282:111965. https://doi.org/10.1016/j.jenvman.2021.111965

Cerdà A, Novara A, Moradi E (2021b) Long-term non-sustainable soil erosion rates and soil compaction in drip-irrigated citrus plantation in Eastern Iberian Peninsula. Sci Total Env 787:147549. https://doi.org/10.1016/j.scitotenv.2021.147549

García-Mollá M, Sanchis-Ibor C, Avellà-Reus L et al (2019) Spain. In: Molle F, Sanchis-Ibor C, Avellà-Reus L (eds) Irrigation in the Mediterranean. Springer International Publishing, Cham, 89–121

García-Ruiz JM, Nadal-Romero E, Lana-Renault N, Beguería S (2013) Erosion in Mediterranean landscapes: changes and future challenges. Geomorphology 198:20–36. https://doi.org/10.1016/j.geomorph.2013.05.023

Jiménez Casero J (2024) Contribución al conocimiento de los mecanismos de colapso de las praderas de angiospermas marinas mediterráneas, Posidonia oceanica y Cymodocea nodosa, bajo los efectos de aportes de nutrientes antrópicos. PhD Thesis, Universidad de Alicante

Kleijn D, Bommarco R, Fijen TPM et al (2019) Ecological intensification: bridging the gap between science and practice. Trends Ecol Evol 34:154–166. https://doi.org/10.1016/j.tree.2018.11.002

Llamas MR, Martínez-Santos P (2005) Intensive groundwater use: a silent revolution that cannot be ignored. Water Sci Technol 51:167–174. https://doi.org/10.2166/wst.2005.0254

López Azorín F (2021) Naturaleza, medio ambiente y repoblación forestal en la Región de Murcia. Fundación Centro de Estudios Históricos e Investigaciones Locales Región de Murcia, Murcia

Lucas-Borja ME, Zema DA, Plaza-Álvarez PA et al (2019) Effects of different land uses (abandoned farmland, intensive agriculture and forest) on soil hydrological properties in Southern Spain. Water 11:503. https://doi.org/10.3390/w11030503

Martínez-Fernández J, Esteve MA (2005) A critical view of the desertification debate in southeastern Spain. Land Degrad Dev 16:529–539. https://doi.org/10.1002/ldr.707

Martínez-Valderrama J, Guirado E, Maestre FT (2020) Unraveling misunderstandings about desertification: the paradoxical case of the Tabernas-Sorbas Basin in Southeast Spain. Land 9:269. https://doi.org/10.3390/land9080269

Martínez-Valderrama J, Olcina J, Gartzia R et al (2023) Uberizing agriculture in drylands: a few enriched, everyone endangered. Water Resour Manag 38:193–214. https://doi.org/10.1007/s11269-023-03663-1

Morugán-Coronado A, García-Orenes F, Cerdà A (2015) Changes in soil microbial activity and physicochemical properties in agricultural soils in Eastern Spain. Span J Soil Sci 5:1304. https://doi.org/10.3232/SJSS.2015.V5.N3.02

Olías M, Basallote MD, Cánovas CR, Pérez-Carral C (2025) Groundwater divide shifting due to pumping in a sector of the Doñana aquifer system (SW Spain): environmental implications. Env Monit Assess 197:526. https://doi.org/10.1007/s10661-025-13965-z

Oosterbaan M, Gómez-Jakobsen F, Barberá GG et al (2025) Characterization and potential causes of a whiting event in the Mar Menor coastal lagoon (Mediterranean, SE Spain). Sci Total Env 978:179391. https://doi.org/10.1016/j.scitotenv.2025.179391

Pemán García J (2024) La restauración forestal en España. Más de 150 años de experiencia. Foresta 89:4–11

Serrano A, Cazcarro I, Martín-Retortillo M, Rodríguez-López G (2024) Europe's orchard: the role of irrigation on the Spanish agricultural production. J Rural Stud 110:103376. https://doi.org/10.1016/j.jrurstud.2024.103376

Sterk G, Boardman J, Verdoodt A (2016) Desertification: history, causes and options for its control. Land Degrad Dev 27:1783–1787. https://doi.org/10.1002/ldr.2525

Van Rijssel SQ, Koorneef GJ, Veen GF(Ciska) et al (2025) Conventional and organic farms with more intensive management have lower soil functionality. Science 388:410–415. https://doi.org/10.1126/science.adr0211

WWF (World Wide Fund for Nature) (2021) El robo del agua: Atlas sobre los derechos del agua en España. WWF España. https://www.wwf.es/el-robo-del-agua (accessed 1 July 2025)

Chapter 17
The Global Trade of Degradation Distorts the Scope of Desertification

Jaime Martínez-Valderrama, Juanma Cintas and Elsa Varela

Abstract The chapter examines how globalization distorts conventional approaches to desertification by decoupling the locations of environmental degradation from those of resource consumption. Traditionally, desertification has been addressed at the national or regional level, with strategies such as Spain's National Action Program to Combat Desertification (NAPCD) focusing on local land use. However, the globalization of food and feed markets has shifted environmental impacts abroad, particularly through the import of soy and cereals for intensive livestock production. While national metrics may show land recovery and reduced overgrazing, these improvements often coincide with the externalization of land degradation to biodiversity-rich ecosystems in South America, where feed crops are cultivated. The chapter also discusses the environmental consequences of Spain's fruit and vegetable exports, including groundwater depletion and pollution, which are linked to international demand. Current frameworks, such as Land Degradation Neutrality (LDN), fail to account for the transnational nature of modern land degradation. Authors call for a reassessment of how desertification is defined and measured in a globalized context, emphasizing the need for international cooperation and responsibility-sharing to address the true scope of land degradation.

Keywords Globalization · Telecoupling · Industrial livestock production · Soybean imports · Land Degradation Neutrality (LDN) · Food trade

J. Martínez-Valderrama (✉) · J. Cintas
Estación Experimental de Zonas Áridas (EEZA), CSIC, Almería, Spain
e-mail: jaimonides@eeza.csic.es

J. Cintas
e-mail: jmcintas@eeza.csic.es

E. Varela
Instituto de Políticas y Bienes Públicos (IPP), CSIC, Madrid, Spain
e-mail: elsa.varela@cchs.csic.es

J. Martínez-Valderrama and J. Olcina Cantos (eds.), *The Labyrinth of Desertification*,
https://doi.org/10.1007/978-3-032-09648-7_17

Spatial Decoupling of Environmental Impact

Interactions between distant regions have always existed, whether economic (e.g., spice trade in the fifteenth century) or ecological (e.g., bird migration across continents). What we now call globalization is an unprecedented acceleration of this rate of exchange. Every action has an impact. Whether we pick a tray of steaks at a supermarket or buy a cotton shirt, we participate in an intricate mechanism where realities disconnected from one another converge. Production processes have become spatially decoupled from their consumption sites, as evidenced by the thousands of kilometers food travels on average to reach our plates (Li et al. 2022).

Desertification is traditionally understood as the result of improper resource use within a specific territory, with the degradation occurring in the same area as the misuse. This notion enabled the delineation of desertification landscapes summarized for the Spanish case (Martínez-Valderrama et al. 2016). All these landscapes are integrated into the NAPCD. These strategies or national plans are the operational units of the United Nations Convention to Combat Desertification (UNCCD). The idea, implemented under the recent paradigm of LDN (Orr et al. 2017; Cowie et al. 2018) is that each country fulfills its respective commitment, resulting in an effective global response to desertification.

However, the phenomenon of globalization, as mentioned at the outset, distorts this assumption. The ratio between domestically produced and imported food has shifted significantly in recent years (Yu et al. 2013). It may not be surprising that this ratio stands at 8–92 for Japan or 20–80 for the United Kingdom-two densely populated island nations. More striking, however, is that in Europe, on average, more food is imported than produced domestically, and that in a country like Spain, which possesses extensive agricultural land, the ratio is 37–63. Beyond national borders, the food we consume generates environmental impacts that are not accounted for in the national balances proposed under the LDN framework.

Ignoring this transnational movement of food and the environmental footprint tied to its production might lead to the false conclusion that a region or country accumulating more biomass and forests is winning the battle against desertification. The cracks in this assumption become evident with the import and export of raw materials and food, revealing severe distortions in implementing the NAPCDs.

The Case of Industrial Livestock Production

The industrialization of livestock production through the concentration of large numbers of animals in industrial-scale farms (so-called "mega-farms") is a clear example of how the environmental footprint of this activity can be effectively obscured. The replacement of extensive livestock systems—based on the use of grazing resources—by intensive systems that rely entirely on imported feed has, in the context of desertification, led to the near disappearance of one of the desertification landscapes identified in the NAPCD—agro-silvo-pastoral systems affected by overgrazing. This has occurred at the expense of a new scenario that only becomes visible when the aforementioned disconnections between places of consumption and production are taken into account.

How is it possible that livestock farming is no longer considered a driver of desertification in Spain, given that the national herd has increased from approximately 9 million livestock units in 1980 to over 26 million in 2020 (Ministerio de Agricultura Pesca y Alimentación 2024a)? This shift is explained by a change in the production model—from extensive to intensive systems—which are largely oriented toward export, as domestic meat consumption has gradually declined from 62 kg per capita in the 1980s (Delgado 2023) to 41 kg in 2023 (Ministerio de Agricultura Pesca y Alimentación 2023). Intensive livestock production, based on confined animal feeding operations, requires the cultivation of cereals and soybeans for feed, rather than relying on pastures and vegetation unsuitable for direct human consumption. Globally, 36% of cereal production (Cassidy et al. 2013) and 75% of soybean production (WWF 2014) are used for animal feed. In total, 6 million km² of land are cultivated to feed livestock. This figure is striking when placed in context: by comparison, 8 million km² are used to produce food for direct human consumption (Ritchie and Roser 2019) (Fig. 17.1).

Within this global dynamic, Spain stands out as the leading producer of compound feed in the European Union (26,479 kt) (FEFAC 2023) and, as of 2023, the second-largest global exporter of pork to China, with exports valued at €800 million. This trade peaked in 2020, reaching €2.534 billion and nearly 3 million tonnes (Ministerio de Agricultura Pesca y Alimentación 2024b). What does this have to do with desertification? To answer this, one must examine the impact of soybean and cereal cultivation used in the production of animal feed consumed in Spain.

Estimates based on soybean import data from South America indicate that, during the first decade of the twenty-first century, approximately 1188 million hectares of

Fig. 17.1 Stall-fed cattle are maintained on a diet of compound feed. Photo by Jaime Martínez-Valderrama

ecologically valuable primary ecosystems were converted into soybean monocultures (Martínez-Valderrama et al. 2021a). The Brazilian Atlantic Forest, the Amazon, the Cerrado, and the Paraguayan and Argentine Chaco are among the main regions affected. These are some of the Earth's most important biodiversity hotspots—formerly impenetrable landscapes where chainsaws and bulldozers have cleared pristine environments to cultivate vast expanses of soybeans.

If our assessment of desertification were limited strictly to Spain, the outlook would appear far more positive. The same study reports that 7078 million hectares of potential grazing lands have improved in condition, as indicated by LDN metrics—namely, an increase in biomass. In short, while the national herd is now the largest in history and forest regeneration is underway, the environmental burden is effectively outsourced to South America's primary forests (Fig. 17.2).

As discussed, not all the meat produced in Spain is consumed domestically; a substantial portion is exported. In the case of pork, China has become Spain's primary trading partner. The outbreak of African swine fever—yet another consequence of the globalization processes in which we are embedded—devastated China's domestic pork production. To meet rising internal demand, China has increasingly relied on imports of Spanish pork, which is produced using feed derived from South American soy. Thus, on the one hand, Spain imports soy and exports land degradation, while on the other, it exports pork and effectively imports the environmental degradation that China would have generated had it raised the pigs itself (Fig. 17.3).

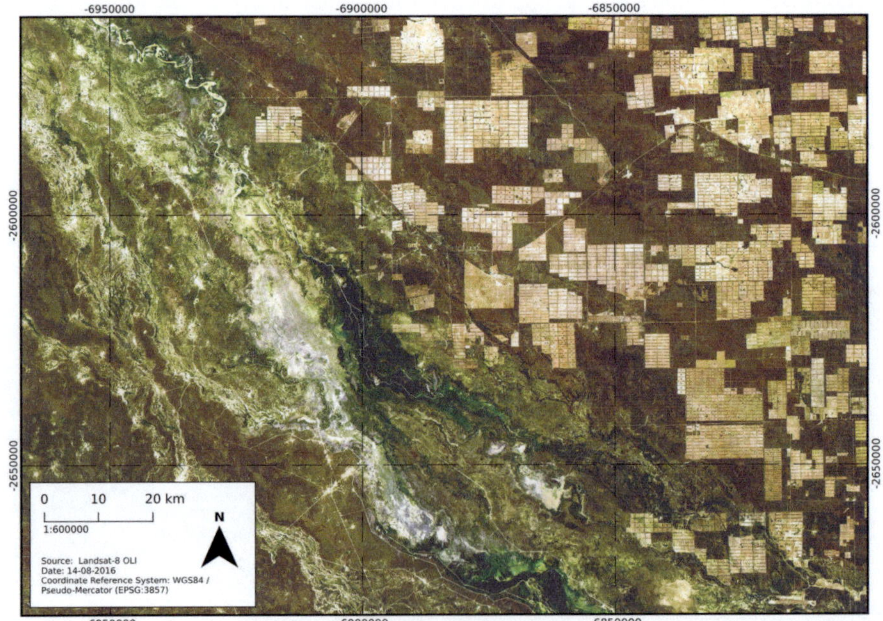

Fig. 17.2 Soybean fields destined for export to Europe and North America have replaced the dry forests of the Chaco region in Argentina and Paraguay, leading to the destruction of ecologically valuable habitats and imposing a significant toll on the region's biodiversity

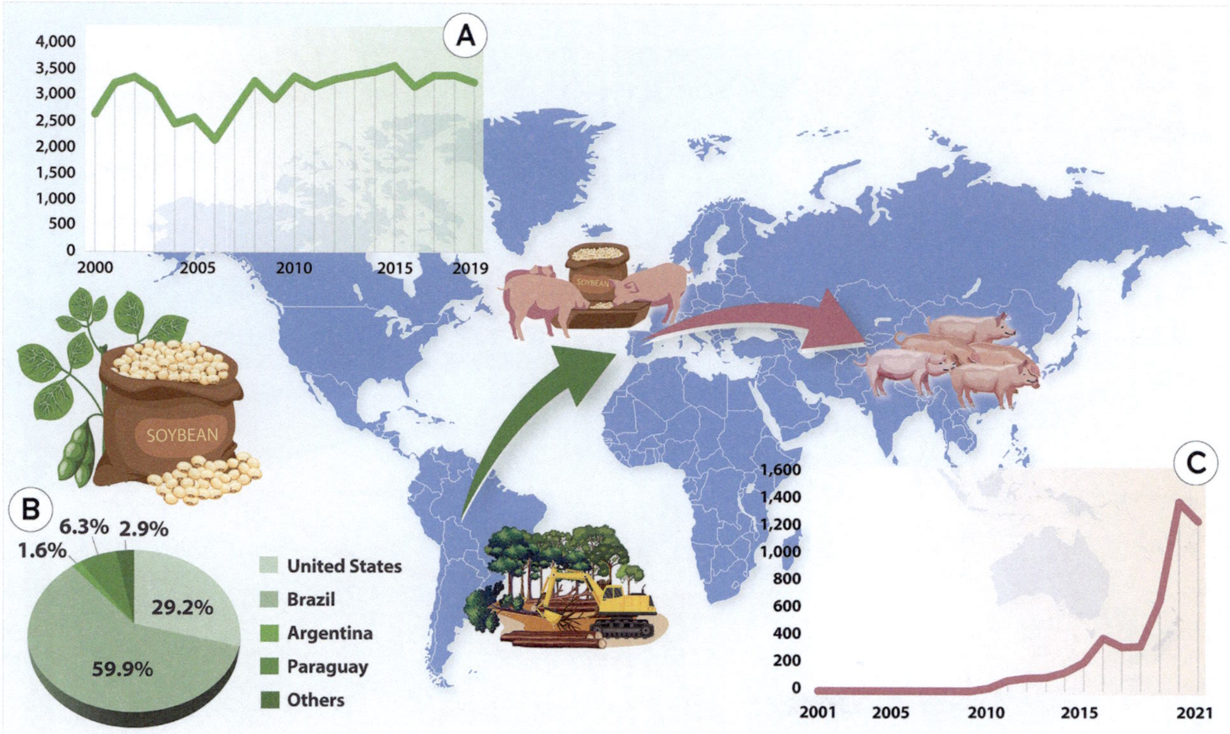

Fig. 17.3 Soybean exports to Spain (**A**, values in thousands of tons; Martínez-Valderrama et al. 2021b) originating from various countries (**B**; Martínez-Valderrama et al. 2021b) are used primarily to feed pigs, which are then exported to China (**C**, values in thousands of tons; European Commission 2025)

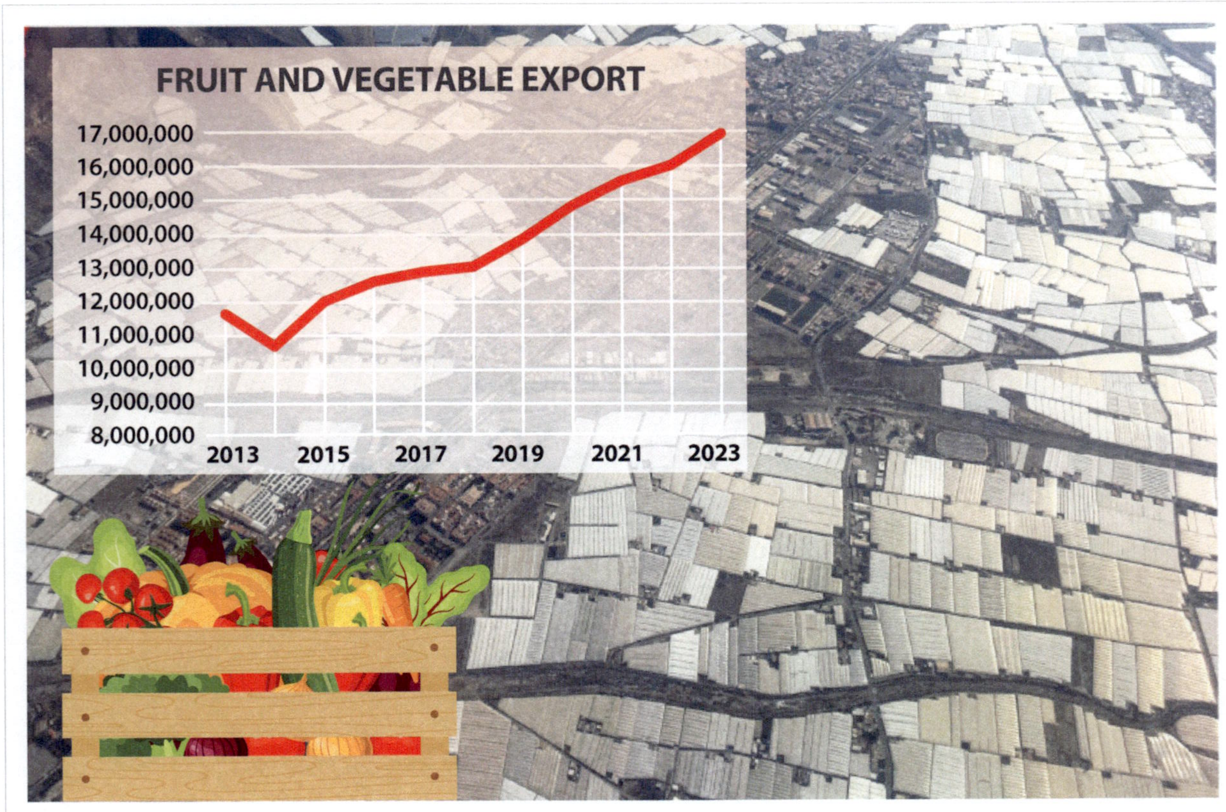

Fig. 17.4 Trends in Spanish fruit and vegetable exports (in thousands of euros, FEPEX 2025), which are mainly destined for European markets. Their production entails significant land degradation

Is the Export of Fruits and Vegetables a Case of Imported Desertification?

Spain hosts the main fruit and vegetable production zone in Europe. Owing to mild temperatures, the virtual absence of winter, over 3000 h of annual sunshine, and the commercial advantages conferred by membership in the European Union, the southeastern Iberian Peninsula has become a major exporter of horticultural products (Fundación Cajamar 2023; Lamo De Espinosa and Garrido 2024).

The environmental side effects of this intensive production—such as the quantitative and qualitative degradation of groundwater resources, and plastic pollution (Martínez-Valderrama et al. 2024)—could arguably be attributed to the countries that purchase these goods. This would, in effect, allow Spain to "offload" a serious desertification issue. However, the commercial benefits of this export-oriented model imply an acceptance of the associated environmental costs (Fig. 17.4).

Who is to "blame"? Assigning causality and responsibility is a contentious issue, yet the location of the degradation is indisputable. Desertification—and more broadly, land degradation (as these processes extend beyond arid zones in the context of regional telecoupling)—must be examined with attention to its international ramifications. Still, it remains unclear how desertification should be geographically allocated in the context of the globalized interconnections that characterize the modern era.

Acknowledgments This research was supported by the ATLAS project (funded by the Biodiversity Foundation of the Ministry for the Ecological Transition and the Demographic Challenge (MITECO) within the framework of the Recovery, Transformation and Resilience Plan (PRTR), financed by the European Union—NextGenerationEU), and by the project "Plan Complementario de I+D+i en el área de Biodiversidad (PCBIO)" (funded by the European Union within the framework of the Recovery, Transformation and Resilience Plan—NextGenerationEU and by the Regional Government of Andalucia).

References

Cassidy ES, West PC, Gerber JS, Foley JA (2013) Redefining agricultural yields: from tonnes to people nourished per hectare. Env Res Lett 8:034015. https://doi.org/10.1088/1748-9326/8/3/034015

Cowie AL, Orr BJ, Castillo Sanchez VM, et al (2018) Land in balance: the scientific conceptual framework for land degradation neutrality. Environ Sci Policy 79:25–35. https://doi.org/10.1016/j.envsci.2017.10.011

Delgado P (2023) From affluence to processed food: meat consumption in Spain from 1950 to the present. Historia Agraria Rev de Agric Historia Rural 91:223–253. https://doi.org/10.26882/histagrar.091e04d

European Commission (2025) Pigmeat trade. https://agridata.ec.europa.eu/extensions/DashboardPigmeat/PigmeatTrade.html (accessed 13 May 2025)

FEFAC (2023) Feed and food 2023. European Feed Manufacturers' Federation, Brussels, Belgium

FEPEX (2025) Exportación/importación españolas de frutas y hortalizas. In: Fed. Española Asoc. Prod. Export. Frutas, Hortalizas, Flores y Plantas vivas. https://www.fepex.es/datos-del-sector/exportacion-importacion-española-frutas-hortalizas (accessed 11 Jun 2025)

Fundación Cajamar (2023) Análisis de campaña hortofrutícola de Almería. Fundación Cajamar, Almería (Spain)

Lamo De Espinosa J, Garrido A (2024) Regadío y Seguridad Alimentaria. La Situación En España. Cajamar, Caja Rural, Almería (Spain)

Li M, Jia N, Lenzen M, et al (2022) Global food-miles account for nearly 20% of total food-systems emissions. Nat Food 3:445–453. https://doi.org/10.1038/s43016-022-00531-w

Martínez-Valderrama J, Gartzia R, Olcina J et al (2024) Uberizing agriculture in drylands: a few enriched, everyone endangered. Water Resour Manage 38:193–214. https://doi.org/10.1007/s11269-023-03663-1

Martínez-Valderrama J, Ibáñez J, Del Barrio G et al (2016) Present and future of desertification in Spain: implementation of a surveillance system to prevent land degradation. Sci Total Environ 563–564:169–178. https://doi.org/10.1016/j.scitotenv.2016.04.065

Martínez-Valderrama J, Sanjuán ME, del Barrio G et al (2021a) Mediterranean landscape re-greening at the expense of South American agricultural expansion. Land 10:204. https://doi.org/10.3390/land10020204

Martinez-Valderrama J, Sanjuán ME, del Barrio G, Guirado E, Ruiz A, Maestre FT (2021b) Data on the re-greening of Spain's landscape at the expense of South American agricultural expansion. figshare. Dataset. https://doi.org/10.6084/m9.figshare.13134977.v1

Ministerio de Agricultura, Pesca y Alimentación (2023) Informe del Consumo Alimentario en España, Madrid, Spain

Ministerio de Agricultura Pesca y Alimentación (2024a) Anuario de Estadística Agroalimentaria (1904–2022), Madrid, Spain

Ministerio de Agricultura Pesca y Alimentación (2024b) Análisis del Comercio Exterior - Informe Bilateral China 2023, Madrid, Spain

Orr BJ, Cowie AL, Castillo Sanchez VM et al (2017) Scientific conceptual framework for land degradation neutrality. A report of the science-policy interface. United Nations Convention to Combat Desertification (UNCCD), Bonn, Germany

Ritchie H, Roser M (2019) Half of the world's habitable land is used for agriculture. In: Our World in Data. https://ourworldindata.org/global-land-for-agriculture (accessed 10 Dec 2024)

WWF (2014) The growth of soy: impacts and solutions. WWF International, Gland (Switzerland)

Yu Y, Feng K, Hubacek K (2013) Tele-connecting local consumption to global land use. Glob Environ Change 23:1178–1186. https://doi.org/10.1016/j.gloenvcha.2013.04.006

Chapter 18
Conclusions

Jaime Martínez-Valderrama and Jorge Olcina Cantos

In *The Burnout Society* by South Korean philosopher Byung-Chul Han (2015), the author presents a fundamental paradigm shift between the twentieth and twenty-first centuries. He argues that we have transitioned from a disciplinary society to a performance-oriented one. This new era is not defined by external repression but by self-imposed pressure to perform, optimize, and achieve. Although Han's analysis does not directly address desertification, the opening of his argument provides a useful framework for our purposes: to understand the origins of a problem in order to address it effectively.

Han contends that traditional societies operated under an immunological paradigm, defining themselves through exclusion—by what was foreign, different, or hostile. Illness came from the outside (viruses, bacteria, enemies), and societies built both physical and ideological defenses to protect themselves. In contrast, today's postmodern, neoliberal society has replaced this model with a neuronal paradigm, where harm is increasingly self-inflicted.

The title of the United Nations Convention to Combat Desertification (UNCCD) fits within this older, immunological logic. The term "combat" suggests an external enemy and frames the issue as something alien to be fought with tools and strategies borrowed from military or defensive thinking. Within this conception, the desert is an invader to be repelled—by building barriers of trees or crops, or by "reclaiming" land through greening efforts. Similarly, drought and aridity are cast as hostile forces to be

J. Martínez-Valderrama (✉)
Estación Experimental de Zonas Áridas (EEZA), CSIC, Almería, Spain
e-mail: jaimonides@eeza.csic.es

J. Olcina Cantos
Department of Regional Geographic Analysis and Physical Geography, Universidad de Alicante, Alicante, Spain
e-mail: jorge.olcina@ua.es

J. Martínez-Valderrama and J. Olcina Cantos (eds.), *The Labyrinth of Desertification*,
https://doi.org/10.1007/978-3-032-09648-7_18

overcome through engineering solutions that counteract nature's injustice—what, in Spain, is often referred to as the "structural water deficit." These approaches reflect an adversarial stance: a confrontation between society and the perceived disorder of nature.

If we adopt the shift in paradigm proposed by Byung-Chul Han, focusing on the internal dynamics of the problem becomes increasingly relevant. While the environment may be more or less conducive to the development of a condition called desertification, it is our management of the land that ultimately predisposes it either to succumb to certain "threats" or to adapt successfully to the ongoing transformations driven by global change, including climate change.

This perspective might help us understand that much of what we label as "water scarcity" may have more to do with rising demand than with an actual decrease in supply. In times of scarcity, nature retreats, waiting for its moment. We, as *Homo sapiens*, do not consider that option. In the face of scarcity, we demand more, generating unsustainable pressure on ecosystems. Perhaps our defining trait as a species is our tendency to confront challenges and demonstrate that human ingenuity can produce yet another miracle to reverse unsustainable trends. Nevertheless, the trail of environmental degradation continues to grow—both in marine and terrestrial ecosystems, the latter being the domain of desertification.

This book aims to emphasize the tremendous complexity of desertification processes, which often begin as phases of economic development. We are not only living in the "age of Fatigue and Burnout," as Han claims, but also in the age of laconism. We are many, and to attract attention, we must express ourselves in ever fewer characters, often resorting to bombastic language. From this arise many of the misunderstandings we carry forward, which hinder the establishment of sound foundations for addressing such challenges.

Desertification is not the spread of deserts—as is commonly seen in the press, with headlines like "The Sahara Desert will invade the Iberian Peninsula," despite how marketable that phrase may be. Nor can it be "combat" (again, a term rooted in the notion of an external enemy) simply by planting trees in extremely arid regions where even shrubs struggle to take root. Rather, desertification is the result of land use practices that are not aligned with the natural characteristics of the environment in which they take place. We must not shy away from explaining what is happening—and from as many perspectives as possible. Only then can we hope to accurately describe the mechanisms driving desertification and intervene where it will be most effective.

Once the problem has been understood in all its complexity—its different facets and possible trajectories—it becomes urgent to focus on solutions. In this regard, the principles of Land Degradation Neutrality (LDN) offer a guiding framework, aiming to stabilize environmental degradation at 2015 levels. Two key elements underpin this approach: first, acknowledging that our way of life contributes to degradation; and second, offsetting that damage through a range of actions. This does not mean merely promoting forest restoration. More fundamentally—and this is the real challenge—it requires us to produce and consume differently.

This may well be the greatest challenge facing contemporary society, as it ultimately involves rethinking our values and preferences. In other words, addressing desertification—as well as the myriad environmental crises we face—demands ethical transformation more than technological revolution. To that end, we must deepen our understanding of how our socio-ecosystems function and recognize the weight of our individual and collective decisions.

This is what our book seeks to support. We hope it contributes, in some small way, to this ambitious goal.

Correction to: The Labyrinth of Desertification

Jaime Martínez-Valderrama and Jorge Olcina Cantos

Correction to:
J. Martínez-Valderrama and J. Olcina Cantos (eds.),
***The Labyrinth of Desertification*,**
https://doi.org/10.1007/978-3-032-09648-7

The original version of the book was previously published without adding the funding information, which has now been corrected. The updated version to the Book is now available with the changes.

Funding Information: "This work was supported by the Biodiversity Foundation of the Ministry for the Ecological Transition and the Demographic Challenge (MITECO) of the Government of Spain, within the framework of the Recovery, Transformation and Resilience Plan (PRTR), funded by the European Union - NextGenerationEU."

The updated version of this book can be found at https://doi.org/10.1007/978-3-032-09648-7

J. Martínez-Valderrama and J. Olcina Cantos (eds.), *The Labyrinth of Desertification*,
https://doi.org/10.1007/978-3-032-09648-7_19

MIX
Papier aus verantwortungsvollen Quellen
Paper from responsible sources
FSC® C105338

If you have any concerns about our products,
you can contact us on
ProductSafety@springernature.com

In case Publisher is established outside the EU,
the EU authorized representative is:
Springer Nature Customer Service Center GmbH
Europaplatz 3, 69115 Heidelberg, Germany

Printed by Libri Plureos GmbH
in Hamburg, Germany